T0285433

Praise for *Sustainability Simplified*

"Antoine de Saint-Exupéry, author of *The Little Prince*, observed, 'If you want to build a ship, don't drum up the men to gather wood, divide the work, and give orders. Instead, teach them to yearn for the vast and endless sea.' *Sustainability Simplified* invites such yearning. It's an appeal that feels aligned to our highest instincts. It's an inspiring 'pull' approach rather than a 'push' strategy. This is the kind of leadership that makes a difference—to inspire people, not just motivate them."

—Stephen M. R. Covey

New York Times and #1 *Wall Street Journal* bestselling author of *The Speed of Trust* and *Trust & Inspire*

"This book is a masterpiece. Regardless of where you fall on the political spectrum—from bluer than Bernie to a proud Trump supporter like me—I'm sure you are concerned about the world we live in. Josh doesn't claim to be on either side of the political spectrum. He's living by his values. *Sustainability Simplified* shows how to improve the environment without having to protest or shout down others."

—Rob Harper

America Out Loud podcast host, author, and political commentator

"*Sustainability Simplified* is about changing mental models. Yes, it's about sustainability—and quite literally saving the planet—but the only way to succeed is to systematically shift our mental models and break constraints. Josh Spodek has lived it and shows us how to. This extremely powerful book will resonate with anyone anywhere on the political spectrum who dares to look at the world, and their life, through fresh lenses. It will unite us."

—Alan Iny

Partner and director, Global Lead for Creativity, Boston Consulting Group

"In *Sustainability Simplified*, Joshua Spodek tells of his beautiful, brave, and personal path to a sustainable life. For himself, and for the planet. He speaks with wisdom, knowledge, and experience. One key takeaway: we will not progress unless we all become intrinsically motivated to be the change we want to see. The Spodek Method described in the book does it. As you will discover, Joshua walks the talk. We don't have to do all he does, but we would do well to follow some. Why? It will make us happier. Separate us from our addictions. And ultimately, heal the planet."

—Guy Spier

CEO, Aquamarine, ardent disciple of Warren Buffett, author of *The Education of a Value Investor*

"Joshua Spodek's story is a wonderful antidote to the hopelessness that so many people—wrongly!—have about the inevitability of our environmental problems. His remarkable tale begins with one man's challenge to avoid harming others. He not only kept at it, but uses his own experience to suggest a bold, convincing vision of the larger transition that our whole society so urgently needs to make."

—Adam Hochschild

Author of *Bury the Chains, King Leopold's Ghost*, and other books; co-founder, *Mother Jones* magazine

"Buckle up and prepare to be deeply challenged! *Sustainability Simplified* presents a refreshingly new and timely perspective on sustainability, spanning addiction to anthropology. Vitally, it offers a route forward, grounded in personal action, step by step, leading to systemic change and ever deeper reconnection with nature and deep joy."

—Andy Samuel

PhD, CBE, non-executive director of UK Met Office, former CEO of the UK North Sea Transition Authority (formerly Oil and Gas Authority)

"*Sustainability Simplified* shows you how to integrate sustainability into your life and inspire those around you, not by changing who you are but by amplifying your existing strengths. The result will be a better world for everyone."

—Mike Michalowicz

Author of *All In* and *Profit First*

"Joshua Spodek has conducted the most remarkable experiment, living in a New York apartment without using any electric power or purchasing any packaged goods or foods in order to live a truly moral life. He discovers that instead of hardship and loss, he has been having a fuller, happier, and more satisfying time. Out of this has come an enthusiasm to help the rest of us get much more out of life by taking much less. Anyone interested in having a better time while doing less harm should read this book."

—Alan Ereira

BBC and independent documentary producer, historian, and founder of the Tairona Heritage Trust benefiting the Kogi

"A truly excellent book. Its arguments about how society changes are unique and compelling, and the Spodek Method is an invitation, not an accusation."

—Lorna Davis

Former CEO of Danone North America (which she established as the world's largest B Corp), former president of Nabisco North America, former president of Kraft China, passionate supporter of the African rhino, and successful TED speaker

"Joshua understands that perception drives actions that change reality. As he demonstrates, we have arrived at a time when changing our perceptions about what it means to be successful humans from ones that emphasize short-term materialistic consumption and profits to ones that promote long-term benefits for all life will lead us to a future that is both sustainable and thriving. This book blazes a trail that takes us toward that future."

—John Perkins

New York Times bestselling author of Confessions of an Economic Hit Man and other books

"Joshua Spodek's Sustainability Simplified is a beautifully written book about a big idea that might just save our planet. Spodek's book is deeply researched and his ideas are expressed clearly and entertainingly. With Spodek's help we're all better placed to take small or even large steps toward leading a more sustainable existence."

—Adam Alter

Professor of marketing and psychology, NYU Stern School of Business, New York Times bestselling author of Irresistible and Anatomy of a Breakthrough

"Joshua Spodek set before himself an arduous task: how to chart a collective transformation of what he calls our PAID culture—the culture of pollution, addiction, imperialism, and depletion. In this boundary-pushing book, Spodek, the quintessential visionary, suggests that we can radically change and adopt lifestyles and a culture of sustainability—but first it will take a revolution in our constitutions, particularly in how we view true abundance."

—Christopher Ketcham

Author of This Land: How Cowboys, Capitalism, and Corruption are Ruining the American West

"Sustainability Simplified has me looking at society completely differently and reconsidering life choices. It is new and different from any book on the environment. Josh helps us understand why we aren't better stewards in our own lives and how to change. Far beyond recycling more or reusing cloth grocery bags, he leads us to deeply examine how we can live richer, more fulfilling lives by dramatically reducing our footprints: avoiding flying, avoiding doof, and supporting a constitutional amendment, to name a few. He shows a clear (still challenging) path to global sustainability."

—Alexandra Paul

Actress, athlete, and award-winning podcast host

"Josh Spodek has been a wonder to me. His commitment to live sustainably, and in so doing, lead the rest of us to do the same, has been fascinating to watch. He is the rare individual who lives by his best beliefs, from morning to night, 365 days a year. There is a lot to learn from Josh and in this book he shares how he did it, and how you can too. Josh will inspire you."

—John Sargent

CEO (retired) of Macmillan Publishers USA

www.amplifypublishinggroup.com

Sustainability Simplified: The Definitive Guide to Understanding and Solving All (Yes, All) Our Environmental Problems

For more information, please contact:
Amplify Publishing, an imprint of Amplify Publishing Group
620 Herndon Parkway, Suite 220
Herndon, VA 20170
info@amplifypublishing.com

Library of Congress Control Number: 2024918407

CPSIA Code: PRV1124A

ISBN-13: 979-8-89138-069-1

Printed in the United States

Sustain ability

SIMPLIFIED

The Definitive Guide to Understanding and Solving All (Yes, All) Our Environmental Problems

Josh Spodek

Contents

INTRODUCTION

Why Did a Construction Worker Give Me $20?

I wasn't always known as the guy who disconnected from the electric grid in Manhattan or who takes five years (and counting) to fill one load of trash at home.

Over eight billion people are alive today and over half of them live in cities. I never expected to be the first I know of among those four billion in cities to take such a bold step. I expected to fail and to have to reconnect my apartment after a few days. I had no practical experience with disconnecting from the power grid. I didn't expect to find nearly everything I had been told about practicing sustainability was backward, including advice by prominent environmentalists.

I was led to expect that steps toward sustainability meant having to sacrifice some of the best parts of life and if too many people tried, we might revert to the Stone Age or a *Mad Max* dystopia. I knew countless reasons to feel discouraged, helpless, and hopeless. Instead, I discovered a vast, global misunderstanding of our environmental problems and therefore no effective solutions, including from prominent environmentalists. No one was doing anything wrong. It's just that as any performer will tell you, until you perform, it's all theory and talk. You have to practice the basics and develop hands-on experience. As far as I know, I became the first of four billion city-dwellers to go meaningfully beyond talking theory to hands-on actions and solutions.

I learned that nearly every environmental issue making headlines and discussed by heads of governments and corporations were *effects*, not *causes*, so their would-be solutions were more distracting than effective. When I uncovered the root causes, which this book reveals, I saw why people feared facing them. They make us feel vulnerable, insecure, and other ways we don't like feeling. But when I did, I found when I implemented actual solutions, they brought results opposite to what I was led to expect: liberation, freedom, community, saving money, saving time, and more benefits. Do such results sound too good to be true? I thought so too. Read on.

One day in June 2022, when I picked up a few pieces of litter in my local park, I heard someone thank me. I looked over to see a guy sitting on a bench wearing a construction vest. Next to him sat a helmet. I guess he was off duty. I wasn't in a hurry so we talked. I shared how I picked up litter daily, but clarified that doing so didn't reduce waste, it only kept it from the ocean a little longer. The greater value of the habit was in helping lead me to buy less stuff that could become litter. I told him how I take years to fill a load of trash.

He asked how. I told him how most of my life I couldn't stop myself from stocking my kitchen with Doritos and Cherry Garcia despite trying and felt guilt and shame. Then I told him how picking up litter made me feel disgust for packaging, which helped me shift to fresh local produce. I no longer needed willpower. I pointed out another benefit: that once I found fresh local produce more delicious, I could eat as much as I wanted of the food I loved and stay fit.

He pointed to his body, which was obese, and said, "I wish I could change, but I'm too far gone."

During the pandemic, the northwest corner of this park had become overridden with heavy drug users. We were within sight of that corner, but not earshot. I pointed to their encampment and asked him, "You know about the people over there who use heavy drugs, right?"

"Yes."

I said, "I've gotten to speak to some of them. They tell me they're too far gone and can't change either."

I paused, then asked, "Can they?"

He looked at me. I could see the gears turning in his mind. "You're right," he said. "I can."

I hadn't said he could stop, so why did he change his mind to say he could? He knew *they* could stop, even if they believed they couldn't. He connected that if they could, he could too.

He got a wad of money from his pocket, peeled off a crisp twenty-dollar bill, and thrust it at me. "Take this," he said.

I didn't understand why he wanted to give me cash. I declined. He insisted. We went back and forth until he told me he would get more from what I said if I took it, so I did, saying I'd give it to charity.

"No," he said. "Spend it on yourself."

In my experience, conversations about the environment or obesity risk becoming emotional minefields, especially with strangers. Yet beyond embracing what I said, he planned to act on it and rewarded me—a stranger he'd never see again—with cash. How did I know what to say? Why do I pick up litter every day? Why did I speak to heavy drug users my neighbors mostly fear and avoid?

My Mission

My goal is not to help construction workers I happen to meet in the park, though I'm happy to when the occasion arises. My definition of leadership is to help people do what they already wanted but haven't figured out how. My mission is to help lead American (and global) culture to restore where people expect that acting more sustainably will bring not deprivation or sacrifice, but liberation, freedom, self-awareness, fun, joy, and other rewarding results.

This book recounts how I noticed this unmet demand for leadership and how to fill it. Since I already taught and coached leadership, I saw how to teach others in experiential courses, workshops, and online communities. They work. Anyone can lead in sustainability. I've made them available online at SpodekMethod.com. People who started out claiming that they knew nothing about sustainability or leadership became inspirational leaders themselves within months.

This book will help you transform in weeks what took me years. You will love the results since they will come from your deepest values. You will feel liberated from oppressive obligations and values others impose on you. You will work hard, and will wish you had started earlier. You will share with others joy, fun, and freedom.

Asleep at the Wheel?

"The vast majority of US companies are asleep at the wheel when it comes to tackling climate change," reported *The Guardian* a decade ago.[1] Last year *The Hill* reported, "DC is asleep at the wheel when it comes to climate."[2] Everyone seems to think someone else is at the wheel, just not acting. Scientists, educators, journalists, activists, and businesspeople act as if: "If I do my part, it will help wake up the people asleep at the wheel. Then they'll act."

Yet no one is steering. Many talk big about sustainability and do things that signal they're acting sustainable, like buying carbon offsets, inadvertently exacerbating the problems they want to solve. Nearly all indicators of our environmental problems continue unabated—deforestation, ocean plastic, greenhouse emissions, extinctions, etc. Environmental catastrophes are compounding, any one of which seems too hard to solve. It's easy to feel overwhelmed. As different as they appear, they all result from our behavior, which results from our stories, role models, images, beliefs, and what makes up our culture. **If we don't change our culture, we can fix problem after problem but we'll end up back here.**

---

1   Jo Confino, "Best Practices in Sustainability: Ford, Starbucks and More," *The Guardian*, April 30, 2014. Links to all footnotes are at SpodekMethod.com/booknotes.

2   Louis Geltman and Shoren Brown, "DC is Asleep at the Wheel When It Comes to Climate," *The Hill*, March 14, 2022.

No profession is designed to develop the skills and experience necessary to change our culture—not science, education, activism, politics, engineering, business, or any of the loudest voices. If everyone only does what they're skilled at, everyone will feel they're doing their part and we will never change course.

No one is asleep at the wheel. No one is *at* **the wheel.** If only there were someone in charge to take responsibility. It would help if a position existed whose role was to fix the environment. It doesn't exist. If we all do our parts, we won't change course, no matter how much we protest, write, vote, donate, innovate technology, tax carbon, or buy more things marketed as "green" or "renewable." Many of those actions will cause us to accelerate the system causing our environmental problems, as we'll see. Our cultural institutions—including Washington DC, Wall Street, Silicon Valley, and academia—propose solutions that, however well-intentioned, accelerate the results of our culture yet more, including its environmental damage. Institutions at the core of a culture trying to change that culture end up like the horror movie when we learn the call is coming from inside the house.

Why Has No One Led So Far?

Why is no one at the wheel? Many people spread facts, numbers, and instructions, but are not *leading* in the field of sustainability. Recall that by lead, I mean to help people do what they already wanted to but haven't figured out how. Why is no one leading in sustainability?

We're a team with a role unfilled. To clarify, consider a team sport like the America's Cup, where boats push the frontier of ship design.

Designers are part of the team. Since they know the ship best, should they also pilot it? No, their skills aren't relevant to piloting. They'd probably capsize. Skippers skipper, not engineers. Everyone does their role best, not each other's job.

So should we expect scientists, teachers, journalists, activists, and politicians to change culture? No, they aren't trained to. Those roles are critical, but we have plenty of people to fulfill them.

We can do what no one has yet: lead in sustainability. We need leaders at every level from global to national, state, and every level to local community, church, workplace, school, and family. Since we can't lead others to live by values we don't, we have to start with leading ourselves. Hence, despite all my training and experience in science, business, education, and more, I focus on developing *sustainability leadership* of yourself and others over science and technology in order to change culture.

This book is for people who want to do all they can, not to blame but to take responsibility and solve our environmental problems. The challenge is as great as any humans have faced. This book will show you what you can do and how your part will make a difference. You will feel called and inspired.

Because you don't live sustainably, you have filled your mind with excuses, rationalizations, and justifications. This book is not a safe space. Consider this introduction your trigger warning. I didn't write to be pointlessly blunt, nor to make you feel comfortable or excuse your complicity. You will have to face and overcome gut checks most people aren't willing to face. I'm going to make you effective and full of meaning and purpose, which requires you to get over yourself so you can act freely and effectively. I started working on sustainability to try to conserve and restore nature. I still do, but my more basic task is to conserve and restore our humanity—especially values embodied in sayings like, "Live and let live," "Do unto others as you would have them do unto you," "Love thy neighbor as thyself," and "Leave it better than you found it," that we don't practice as we used to.

Some may use this book as self-help to stop procrastinating on changing behavior they know hurts others, personally thriving through living sustainably. If you just use it to improve your life, great! Many, however, will choose to *lead*. This book and my work are designed to make leaders who love rolling up their sleeves, working hard, building community, inspiring others, and making more leaders. I wrote this book to create great, historical leaders.

To change culture toward sustainability requires leadership (helping people do what they already want to do but haven't figured out how). Most people understand that to lead in sustainability you need to know both how to lead and the science sustainability involves (which can include deep knowledge of nature from living in it; many members of indigenous cultures know more relevant science than "scientists"). A scientist trying to lead without leadership experience tends to lecture but not motivate or inspire. A leader without science experience tends to promote ideas that sound nice but don't work or are counterproductive, like carbon offsets. Our world contains many scientists and many leaders, but nearly none experienced as both.

Few recognize that sustainability leadership requires experience living sustainably, or at least trying. Without this third dimension, would-be sustainability leaders are like music teachers who haven't played an instrument. They can teach music appreciation but not how to play. To lead sustainability with integrity, credibility, character, and other properties of effective leadership requires all three dimensions.

Why You, Here, Now?

Meanwhile, you have to put food on the table. You likely have rent or a mortgage to pay, a retirement to plan for, and maybe kids to feed and put through school. Or maybe you're in school yourself or haven't started your career, and you're looking at the seven or eight decades you expect to live. You didn't ask to be born into a polluted, depleted world or a culture that makes stopping polluting and depleting hard. (By pollution, I mean hurting people or wildlife through releasing harmful substances into the biosphere. By depletion, I mean taking from nature faster than nature can replenish, depriving future generations of equal access.)

Most of us just want to live our lives. We didn't cause the problem and we don't want to make it worse. On the contrary, I bet you would like to make it better. Perhaps you love nature. You probably don't like meaningless work but don't shy away from meaningful and effective work even when it's hard, like raising kids, contributing to your communities, and learning in school.

Yet front-page news of environmental problems assaults us nearly daily. When I grew up, such stories appeared a few times a year. They'll become daily soon and more grave all the time, along with news of wars, famine, disease, and other catastrophes. They keep reporting problems we can't see how we can help solve. What exactly are we supposed to do here and now on carbon taxes or international treaties? Actionable things like buying carbon offsets and recycling keep getting debunked as ineffective. Does anyone really expect us to think avoiding straws, flying, or meat will fix things globally? Or even choosing to have a smaller family?

Whether they come from scientists, teachers, journalists, activists, or politicians, these stories make us feel helpless, hopeless, guilty, fearful, ashamed, and full of despair. And the voices keep repeating themselves, ever more shrill, as if calling our problems "emergencies" and "crises" will help. They just make us feel more anxious.

Besides, none of these people are living the way they want us to. The most prominent seem to be making money, gaining power, and enjoying the fame that comes from scaring us. Meanwhile, they don't do what they recommend, do fly private jets to environmental conferences, and more. If they believed what they said, they would do what they recommend. They don't, so what else can we do but follow the money and conclude they're self-serving?

They're why I didn't act for decades. With a PhD in physics, I felt I understood the science as well as anyone. Having co-founded a company operating on four continents, run it as CEO, and earned an MBA, I knew how to create teams and implement solutions. Yet the news made me feel anything I could do wouldn't matter, that only governments and corporations could make a difference on the scale we need, and affirmed all the other reasons not to act we've all felt.

I knew my polluting and depleting were contributing to the problems yet saw no way to help, so I felt as hopeless and anxious as anyone. The biggest value of the years of work leading to disconnecting from the grid and coaching construction workers in the park was that they forced me to face every obstacle and figure out how to overcome them all.

The value of my prior experience leading, coaching, and training others to lead was knowing how to help others implement in their lives what I did in mine. I don't suggest others copy my personal activities. Everyone's situations are unique, so how they implement solutions often looks different than mine, but they work for them in their lives.

I only started my sustainability leadership work when I saw I could help others and not just tell them what to do. I only started spreading the word when they responded with gratitude and wanted to bring to their communities what I brought to them. I promote my work not to support flying to climate conferences but because we breathe the same air and because people feel anxiety over what could bring them liberation and joy.

You wouldn't believe the friction, misunderstanding, and accusations I faced at every stage, and I'm not just talking about from my mom and dad. It came from every part of society. Developing a leadership technique on sustainability was like trying to cross an emotional minefield, where each person's emotional mines were in different places and I had to start over after hitting each landmine. I had to cross over and over, hitting mine after mine, until I learned how to cross, bringing people with me.

With experience, I boiled the technique to its essence: a mindset shift that I could produce consistently, reliably, and predictably. Without it, people pushed back. I saw that I had provoked that resistance, as did the scientists, teachers, journalists, activists, and politicians before me. We were causing the resistance we wanted to overcome.

With the technique, the mindset shift led to a process of continual improvement. People see and feel that since they experience reward from acting more sustainably once, they will do so again, as will others. That is, together, we will create a self-sustaining, accelerating movement.

The technique became known as the Spodek Method—a sustainability leadership technique combining my experience teaching and coaching leadership with my experience practicing sustainability. It activates people's *intrinsic* motivations relevant to sustainability and emphatically avoids motivating through *extrinsic* motivations or lecturing facts and numbers. I envision it playing a role in sustainability like that of civil disobedience in India's independence and America's civil rights. It is a tactic that enables a strategy to achieve a mission.

Our environmental problems may not be our fault, but they are our responsibility if we want to solve them. When people who learn the Spodek Method practice with others and hear people they led do more than they committed to, they want to share their results and teach more. When others show gratitude for being led to act more sustainably, they realize it is different than anything on sustainability they've seen or heard. Some train to become workshop leaders and start initiatives on their own. Some get promoted to sustainability positions in their firms when their bosses see their genuine, authentic, and effective activities that "sustainability professionals" rarely do.

You're already reading this book. For you to act, all you have to do is finish it. If you find it compelling, you can learn to practice the Spodek Method through the how-to workbook, a free download from SpodekMethod.com. I'll be candid: of the thousands of people I've led through the course, it didn't work with a few, but with the overwhelming majority, it did. I predict you'll experience that mindset shift, enjoy the continual improvement, and want to share it with others. You can also

The Surprisingly Joyful Path

Despite popular misconception, living sustainably is joyful, when led there effectively, which the Spodek Method does. This sequence shows some of the stages you will go through, and will lead others through.

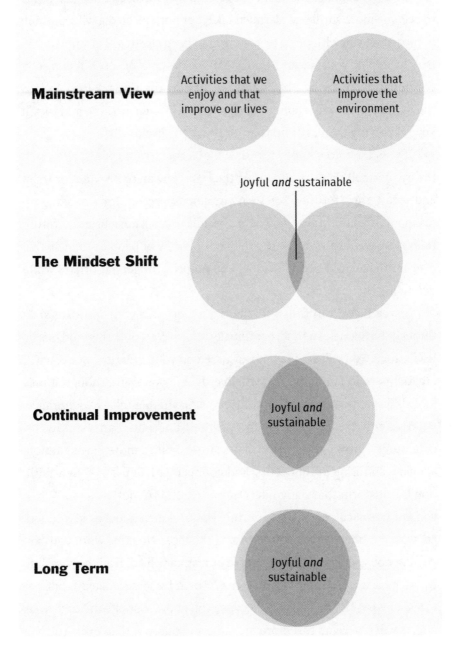

Mainstream View

Activities that we enjoy and that improve our lives

Activities that improve the environment

The Mindset Shift

Joyful *and* sustainable

Continual Improvement

Joyful *and* sustainable

Long Term

Joyful *and* sustainable

find webinars, testimonials, and hundreds of episodes of my podcast, *This Sustainable Life.*

In the meantime, the media is dominated by people whose best response to people dismissing their attempts to shock and outrage you is to become more shrill and alarmist. Like it or not, the media will continue bombarding you with news to scare you to keep clicking. They may think they're helping you, but until you experience your mindset shift and start continually improving, they will keep sinking their hooks into you. In their defense, the environmental problems will keep growing and they'll keep reporting them. It's up to us to break the cycle.

If you don't do your part, you won't be able to look your children in the eye and tell them you're doing the best you can to provide for them and secure their futures. You won't be able to prepare for a relaxing retirement. If you're young, you won't be able to envision a brighter future to look forward to. Even in the midst of billions of people, you will feel ever more isolated because all will be pointing fingers at others while defending themselves.

Instead, you can finish this book, experience your mindset shift through the Spodek Method, continually improve, enjoy acting, and maybe lead others. You will act in your way based on your situation, but you will create liberation and joy. You will live the life you want, though it may look different than past generations projected in their blissful ignorance.

You will create security for yourself and your family, develop community, grow nature, appreciate nature, build a more perfect union, alleviate suffering of innocents, and connect with and serve your faith. You'll transform that daily news from dreadful to motivational. When you know you can alleviate suffering, you prefer to know of it. You will act with confidence that you are doing the most effective work you can.

I'm not saying we'll create utopia or manna will fall from heaven. Life will require work. People will disagree. War and disease will exist, but not global environmental collapse. Because you know your efforts work, your efforts will make you feel more meaning and purpose than ever. You will replace isolation with community, despair with enthusiasm, and passivity with determination and leadership.

All of us are called. You are needed. Your specific history, connections, skills, and experiences will help in ways no one else's can. You will receive help in ways you never expected. You will not measure the quality of your life by things outside your control but by whether you are reaching your potential to improve your life. Also the lives of those you love and who love you, your community, your nation, and innocent people globally. You will feel gratitude and enthusiasm for living when humanity is facing one of its greatest challenges. You will love the meaning and purpose in rising to the challenge and doing through teamwork the best you possibly could.

Government, Policy, Liberty, and Freedom

The Spodek Method goes beyond the personal to policy and the root of the role of government. Maybe you prefer smaller government and free markets that allow innovators to solve problems. Pollution destroys life, liberty, and property without the consent of the people deprived of their lives, liberty, and property. Instead of protecting these things, nearly every government in the world profits from permitting that destruction and therefore promotes it. When I describe our markets as being coercive, I mean when government permits and even promotes the destruction of life, liberty, and property, including activities that pollute or deplete.

In particular, if in creating a product, I put toxins into the environment, I hurt people without their consent. I coerce them into accepting my pollution. A free market means trade is voluntary and therefore mutually beneficial. Ideally, free markets allocate resources to those who solve problems best.

A coercive market is not mutually beneficial. No one consents to being born with birth defects or getting cancer. Coercive markets may allocate some resources to those who solve problems, but they also allocate resources to those who coerce the most, enabling them to coerce yet more, making a mockery of free markets.

Government generating revenue and power from taking or destroying what it is designed to protect is a conflict of interest, if not a moral hazard. Just permitting that destruction by others leads government and market solutions to increase government size and ability to coerce.

Yet few oppose government's essential role in protecting life, liberty, and property. Nobel laureate in economics Milton Friedman, no fan of big government, said "I'm not in favor of no government. You do need a government . . . There's no other institution in my opinion that can provide us with protection of our life and liberty." He knew that "the key insight of Adam Smith's *Wealth of Nations* is misleadingly simple: if an exchange between two parties is voluntary, it will not take place unless both believe they will benefit from it. Most economic fallacies derive from the neglect of this simple insight."

This book, and doing the Spodek Method, shows how to extricate government from profiting from the destruction of life, liberty, and property. Doing so may be challenging, but defending freedom usually is. The result will be to restore freedom, liberty, innovation, entrepreneurship, and leadership in nations who do so first.

My Vision

I envision a world of vibrant culture, full of entrepreneurship and innovation but not pollution or depletion. We will enjoy high quality of life, health, security, community, and longevity. I am a huge fan of technology, though I believe government should protect you from my using it in ways that deprive you of life, liberty, or property. I foresee this book's recommendations leading to more innovation than ever, creating more liberty than ever.

This book will present a way there through new perspectives, some new information, an experiential technique, and a WWII-level mobilization project. The project will be challenging to envision but will be as effective as past generations' mobilizations to fight Hitler, eradicate smallpox, and walk on the moon. It will sound challenging, but in time will reveal itself as doable and desirable. Best of all, it will appeal to your deepest values, no matter your politics or identity.

Conservatives and libertarians may be pleased to find this project is based in traditional views including Enlightenment values and American exceptionalism, as expressed by voices including Edmund Burke, Friedrich Hayek, Russell Kirk, Frédéric Bastiat, and Ronald Reagan. It promotes free trade, free markets, innovation, entrepreneurship, traditional family values, national security, and small-town America. It opposes centralized planning, big government, woke-ism, handouts, and more. It exposes how coercive and unfree our markets are, how unsustainability bloats and corrupts government, and undermines traditional values. I am surprised at how readily conservatives and libertarians are embracing and practicing in the area of the environment coercion, central planning, and overriding classical liberal tradition. I am surprised how much I see libertarians supporting practices that lead to government permitting people to destroy others' life, liberty, and property, even profiting from it.

At the same time, liberals may be pleased to find this project is based in indigenous wisdom and global diversity. It promotes social justice, equality, and fairness. It fights racism, sexism, imperialism, colonialism, speciesism, and more. It exposes how much our culture leads you and me to fund domination and injustice. I am surprised at how much I see liberals embracing environmental policies that take from innocent and helpless people and give to the rich and powerful. They champion caring for others in words, but their behavior undermines that caring.

My Bias

Once I participated on a panel discussion. I'm sure you've attended ones like it: four panelists were on the stage with a moderator. After we spoke, the moderator took questions from the audience.

Before we could answer one of those questions—I don't remember the exact question—the moderator said to the audience member, "You asked a general question, maybe figuring that the more general the question, the more people it would resonate with. It turns out it works the other way. People connect more with details, even if your details differ from theirs. Can you ask your question again, but relevant to a specific problem in your life you want to solve?"

The advice worked. The audience member rephrased the question to ask about a particular part of their life. I felt more engaged and sensed the room did too. For that reason, I share many personal stories. I don't want you to copy my actions. I want to help you live by your environmental values. They will look in your life different than mine in my life, but you can implement yours as much. You will find as much meaning and reward. Many of you will do more than I have, however different it may look.

Also, while our environment and its problems are global, I will focus on the United States—not to exclude other places, but to engage with details that people elsewhere will be able to act on more effectively than if I spoke in generalities.

I know American culture best so can be more specific about it. America leading the world in unsustainability makes us arguably the most important nation to change, and in many ways the most challenging and resistant. I expect that solutions that can work here will work more and faster elsewhere.

Conversation in the US about sustainability tends to be tribal: liberal versus conservative versus libertarian, old versus young, rich versus poor, and so on. Yet we all want clean, pure air, water, land, and food. We want to live free from coercion. At times I considered writing multiple versions of this book since when I wrote in the language of one tribe, other tribes disengaged.

I could only write this book after finding a solution common to all— hence sustainability *simplified* in the title. I was satisfied and motivated by seeing how understanding the past led to a solution consistent with the values of Milton Friedman, Barry Goldwater, Ronald Reagan, Ayn Rand, and other conservatives and libertarians as much as Rachel Carson,

Ralph Nader, E. O. Wilson, Jane Jacobs, and other liberals; as well as the San, Hadza, Kogi, Iroquois, and other indigenous cultures.

Milton Friedman
Ronald Reagan
Barry Goldwater
Ayn Rand

Rachel Carson   San
Ralph Nader   Hadza
E.O. Wilson   Kogi
Jane Jacobs   Iroquois

An Overview of This Book

In this book I will help you understand our situation so you can act to help yourself, your family, your community, your nation, and our world. The Spodek Method helps you do what you want but may not have figured out how.

First, what this book is not. I am not trying to convince, cajole, or coerce; nor judge or shame; nor lead by example. I have researched and debated extensively and have concluded our environmental problems exist and are among the most important we face, but I did not write to lecture, argue, or debate. I wrote this book for people who want to act but haven't figured out what to do. If you want to argue or debate, this book isn't for you. I cite in the coda my top sources I found compelling as well as the top ones I studied but found uncompelling. If you think I'm missing something, I welcome new views and would love to change mine for compelling reasons. If you contact me, though, please make sure you aren't repeating what I listed. Still, people, corporations, and governments aren't resisting for lack of facts, numbers, or instruction on sustainability, so repeating them doesn't help, nor does insisting on calling our situation a "crisis" or an "emergency."

That said, I do discuss some science, but not by saying, "Here's the science. You must agree with me and therefore you must do what I say." Humans are subject to cognitive biases. We have amazing abilities to believe something we want or disbelieve what we don't. For example, I desperately clung to beliefs that energy sources like solar, wind, nuclear, and, if it ever worked, fusion were "clean," "green," "renewable," and "emissions-free," as many claimed. I believed those characterizations partly because they sounded scientific and I knew more science than most. Looking back, though, I see that I believed them more because I felt I would lose hope without them. I feared that if markets or population didn't grow, they might collapse. It turns out I was wrong, but could only accept the evidence showing so after my mindset shift. It transformed what looked like a bitter pill to swallow into giving me comfort and confidence not to rely on wishful thinking. I refer to the relevant resources, though experience has shown that, like me, people only accept them after our mindset shifts. I also refer to resources that show human ingenuity's ability to overcome the challenges of foregoing energy sources we thought we couldn't live without.

Part 1 describes why you have every reason to be confused into inaction on the environment despite how much you care. Few people have experience in all three necessary areas of leadership, science, and living sustainably. No one else knows what they're talking about regarding a full solution. They don't know that they don't know what they're talking about. Regarding the problems, they tend to understand the symptoms but not the causes. I also share my background. Some might know me as the guy who unplugged from the grid in Manhattan, but I didn't just disconnect the circuit. The step of disconnecting was a small one in a series of simple, small, liberating experiments. Until those experiments, I had given up and struggled with fears, insecurities, failure, and vulnerability. I share this journey.

Part 2 describes the world we were born into: a world of pollution, depletion, addiction, and coercion. I'll clarify whom I mean by "we." Our largest, most profitable corporations and governments have mastered manipulating our motivational systems often more effectively than we can resist. I show what happens to our minds and culture when we act against our values. I describe the psychology of corruption and the resulting loss of reason. Our environmental problems result more from (and cause) loss of reason, freedom, and democracy than lack of innovation or carbon taxes.

Part 3 describes how we got here. I share how 250,000 years of human existence led to this culture. I share how the anti-democratic forces that dominate us came to exist and grew to undermine our reason, freedom, and democracy. There aren't bad guys, just billions of people procrastinating, including the people you think should be acting. You'll come to see those we blame with more compassion. We will see ourselves in them. There aren't good guys either, which is why no one is asleep at the wheel. I show examples of people and movements that have changed cultures on the scale of nations, empires, and the globe before. They are role models.

Part 4 describes the solution. Since our problems are rooted in loss of freedom, reason, and democracy, effective solutions are based in restoring them. We know humans can live without destroying each other's life, liberty, property, and freedom, which polluting and depleting do, because our ancestors did for 250,000 years. Nearly everyone sees the global scale of our environmental problems and limited time to act. Many recognize the need to mobilize at greater scale and speed than we did to fight Hitler, eradicate smallpox and polio, and reach the moon. We succeeded at those things. They took effort, but people involved loved the work. I will present our WWII-level mobilization project that we will love—a stewardship version of the Thirteenth Amendment. It will at first look impossible, then desirable. Soon, we will achieve it.

An epilogue describes the feeling of acting on this book and a coda points to the literature I found most valuable and readable.

Online and In-Person Resources

I've created resources for you to learn the Spodek Method experientially. They include a how-to workbook for free download, online courses, online community, tools to find and create in-person communities, and more.

Living sustainably isn't hard. Living sustainably when nearly everyone around you doesn't is. These resources and communities are like an oasis of support and platform to excel. They are available at SpodekMethod.com.

I couldn't describe the results of hands-on experience based on Spodek Method workshops better than Beth 🎙 [3], who shared with me the following email she sent to friends:

> I would like to share with you my experience with confronting climate change head on this year. I decided to make it the year I stop my gloom and doom and to let go of my self-talk that reinforced that I am helpless to do anything. I am discovering that changing my own behavior is joyful and empowering. Deprivation and sacrifice are the OPPOSITE of how I feel about the daily journey toward habits that care for our beautiful planetary home.
>
> How did I come to this change of heart? My daughter took a class with Josh Spodek in Sustainability Leadership and I happened to be at her house while she was taking it. This led to conversations that challenged my pessimism about being able to do anything more than I was already doing. My pessimism about individual action making any difference was challenged. It fundamentally came down to "I can continue along as I am and for certain nothing will change, or I can take the reins of my part of this giant puzzle and have the chance to be a part of the solution."
>
> A large part of my motivation came when I used an online carbon calculator to determine my "carbon footprint." I discovered that

---

3   The microphone symbol after someone's name indicates that they were a guest on my podcast. All episodes are available at http://joshuaspodek.com/all-podcast.

from flying alone for the first seven months of 2023 I had belched out over 10 times the amount of carbon that is considered the "sustainable limit" per person per year. This number didn't even include gasoline, natural gas, or any other modes of consuming or polluting. It literally made me cry. It also made me get serious.

I took the course that my daughter had taken and found a source of support, inspiration, information, and skills that were new. One of the things about this class that I think is most powerful is that there is nothing "prescriptive" about it. There are no lists of things you should do now and things you should avoid now. No one is deciding for you or shaming you into choices. Instead, it is an inward journey of connection to one's own internal motivation that is grounded in our own experiences in nature. It is a process of continuous improvement, so I didn't decide to reduce my trash consumption and then stop when I did that. I look every day for new ways to lessen my impact, and every time I find another way I feel GREAT and motivated to figure out what's next.

I am writing to invite you to take this class. Josh's model is to use conversations with each other as the foundation of connecting to our internal motivation, conversations using the Spodek Method. These conversations help build a community of people who have experienced the joy of taking self-directed action in one's own life. As with any BIG problem, the solutions require all of us. This class helps, one person at a time, to build a community of people who see themselves as part of the solution. I think you will be surprised and delighted with the empowerment you feel to take action.

PART 1

WHY SUSTAINABILITY LOOKS HARD AND WHY NOT TO BE DISCOURAGED

CHAPTER 1

OVERLOAD
AND CONFUSION

The United Nations may be the world's greatest voice *discouraging* environmental action, promoting *inaction*, and even accelerating the environmental problems it warns about.

"Wait, isn't the UN supposed to help?" you might ask.

Despite its intent, the UN confuses, overwhelms, and discourages acting more sustainably. First it says *CODE RED! BILLIONS OF LIVES IN DANGER! ALL MUST ACT NOW!* Next it contradicts itself, saying people won't die. Then it doesn't act itself. Let's look at the details.

Its first message comes from the Intergovernmental Panel on Climate Change, which it created. The Secretary-General António Guterres announced the panel's results: "CODE RED!", which made global headlines. A quick search found "Code red for humanity" headlines in *The New York Times, Forbes, The Guardian, Financial Times, Reuters, Associated Press, PBS, The Independent, BBC, Bloomberg, Voice of America, NPR, The Washington Post, Al Jazeera, USA Today, Fox News, Breitbart, CNN*, and more. He elaborated, "Greenhouse-gas emissions from fossil-fuel burning and deforestation are choking our planet and putting billions of people at immediate risk."[4]

The second message contradicts the first. The UN's population

---

[4] "Secretary-General Calls Latest IPCC Climate Report 'Code Red for Humanity,' Stressing 'Irrefutable' Evidence of Human Influence," United Nations, August 9, 2021.

predictions show the population smoothly leveling off around ten billion people, then modestly decreasing around 2100.[5]

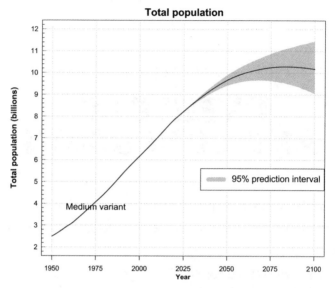

Total population

© 2022 United Nations, DESA, Population Division. Licensed under Creative Commons license CC BY 3.0 IGO. United Nations, DESA, Population Division. *World Population Prospects 2022.* http://population.un.org/wpp/

Whatever the danger, the smoothness of that curve sure looks like no one will die from it. We also seem to have a comfortable cushion since the shaded region says the population might also increase, implying the UN believes the earth could sustain over twelve billion people.

What do we conclude from these two messages? Well, if the UN says that everyone is in danger but it won't lower population levels, then the question becomes how we will feed and house ten billion people. We know rich people will have enough, so the UN implies we should produce enough to reach the poor. In other words, the UN implies we should *increase* production—thereby polluting and depleting more.

Now the third message. Guterres also said "the era of global warming has ended; the era of global boiling has arrived. Leaders must lead. No more

5   "World: Total Population," United Nations, 2022.

hesitancy. No more excuses. No more waiting for others to move first."[6] What does UN leadership look like? Here is a typical month of Guterres's travel:[7]

- Qatar, 01 May to 02
- Kenya, 02 May to 05
- Burundi, 05 May to 06
- Spain, 07 May to 09
- Washington DC, 10 May
- Jamaica, 14 May to 15
- Japan, 18 May to 21

Does flying to seven countries on four continents per month look like he is following his own advice? Not to me. You could argue his travel helps the world more than it hurts, but that claim, if left unexplained, conflicts with, "Leaders must lead. No more hesitancy. No more excuses. No more waiting for others to move first." The UN claims carbon neutrality, but those claims do not appear to withstand scrutiny.[8,9] The conflict between his words and behavior leads people to feel: "if he flies around like that . . . well, my work and family are important too. I guess I should fly as much." Could the UN with its resources lead by polluting and depleting less, say, by delegating these visits and empowering local leaders? He might be more effective polluting and depleting less.

What is the UN's behavior saying if not: "if you can justify it, keep polluting and depleting"?

I focused on the UN, but you'll find similar patterns in nearly every other group and person promoting sustainability, including Al Gore,

---

6   António Guterres, "Secretary-General's Opening Remarks at Press Conference on Climate," United Nations, July 27, 2023.

7   "Official Travels of the Secretary-General."

8   Lisa Song and Paula Moura, "An (Even More) Inconvenient Truth: Why Carbon Credits for Forest Preservation May Be Worse Than Nothing," *ProPublica*, May 22, 2019.

9   Jacob Goldberg, Léopald Salzenstein, Sarah Brown, and Shaz Syed, "Is the UN Really Climate Neutral? No," *The New Humanitarian*, September 13, 2023.

Extinction Rebellion, 350.org, Silicon Valley, countless NGOs, the mainstream media, and the US Democratic party. I'm not letting voices off the hook that promote activities that pollute and deplete like extractive industries, the US Republican party, libertarians, industrial agriculture, and their peers. I address them violating their values after covering the problems of loss of freedom, reason, and democracy.

Others Say, "No Problem. Things Are Getting Better."

Meanwhile others suggest we're in the best times ever. By many measures of the economy, health, longevity, and comfort, we are. They aren't saying one thing and doing another, so they speak confidently.

Perhaps most representative is Harvard psychologist Steven Pinker in his books *Better Angels of Our Nature* (2011), which Bill Gates described as "the most inspiring book I've ever read," and *Enlightenment Now* (2018), which Gates described as "not only the best book Pinker's ever written. It's my new favorite book of all time."

In these books, he claimed that while wars and violence still happen, we live in the most peaceful time in history and that today's peace results from the Enlightenment values of "reason, science, humanism, and progress." I began with skepticism but found he backed these claims up conclusively. Regarding the environment, he acknowledged problems, but promoted a movement called "ecomodernism," that promotes applying those Enlightenment values to keep solving problem after problem. He writes, "Industrialization has been good for humanity. It has fed billions, doubled life spans, slashed extreme poverty, and, by replacing muscle with machinery, made it easier to end slavery, emancipate women, and educate children. It has allowed people to read at night, live where they want, stay warm in winter, see the world, and multiply human contact." He goes on to state that "the tradeoff that pits human well-being against environmental damage can be renegotiated by technology."

He didn't start ecomodernism. The ecomodernists wrote a manifesto echoing that technology would "decouple" the desired outcomes of technology from the undesired effects like pollution.[10] Peer communities suggested "green growth" and a "circular economy" would achieve similar goals. Economist Julian Simon's 1998 book *The Ultimate Resource 2* said that if we run out of any material, its scarcity would cause prices to rise and we could do what we always have: substitute other materials, dematerialize, and so on. Ideas from human ingenuity solve problems and ideas are in infinite supply. The more people, the more ideas, so we should keep growing the market and population. Marian L. Tupy and Gale L. Pooley's 2022 book *Superabundance* followed up with more data showing that more growth correlates with improved quality of life. Economist Milton Friedman said that free markets allocate resources to those innovators best able to solve problems so we shouldn't interfere with them. Alex Epstein's bestselling *The Moral Case for Fossil Fuels* suggested that, whatever problems fossil fuels cause, they enable solving our problems as well, so we should keep using them.

Pinker described those crying "Code Red" as seeing using energy "as a heinous crime against nature, which will exact dreadful justice in the form of resource wars, poisoned air and water, and civilization-ending climate change. Our only salvation is to repent, repudiate technology and economic growth, and to revert to a simpler and more natural way of life." They are a "quasi-religious ideology . . . laced with misanthropy, including an indifference to starvation, an indulgence in ghoulish fantasies of a depopulated planet, and Nazi-like comparisons of human beings to vermin, pathogens, and cancer."

I like peace, prosperity, and Enlightenment values like reason, science, humanism, and progress. I'll call them Enlightenment values, but I'll note they didn't start with the Enlightenment—they were just restored after being mostly lost in Europe for centuries. Our ancestors practiced them

---

10   John Asafu-Adjaye, Linus Blomqvist, Stewart Brand et al., "The Ecomodernist Manifesto," Ecomodernism.org, April 2015.

long before agriculture. Should we just keep letting them work their magic?

As it turns out, there is another Enlightenment value that Enlightenment thinkers recognized as essential but our culture has lapsed in practicing: stewardship. The consequences of its lapse were too small to notice until recently, but its absence undermines the predictions of Pinker and his peers of continued peace. As we'll see, a lack of stewardship and sustainability will always undermine freedom, reason, equality, and democracy.

A big part of my sustainability journey was identifying that I grew up in a culture neglecting stewardship, the globally deadly consequences of its absence, and how to restore it. Part 4 will show how stewardship—that is, living in ways that enable and ensure future generations can enjoy the world you do—was both essential to the Enlightenment and rooted in indigenous practice and wisdom. That universality tells me Pinker, his peers, and people of all political stripes, ages, backgrounds, and identities will agree on its value and that we should restore it.

Do We Lack Political Will?

You've doubtless heard the sentiment "We Have the Technology to Solve Climate Change. What We Need Is Political Will" as declared a recent headline, but this view misunderstands our situation.[11] Technology is a red herring. Saying we need "political will" implies voters are asking for something politicians aren't delivering. Meanwhile, any politician can see voters spend their money on SUVs, bottled water, and so on. Protesters may march once a year or so chanting "There Is No Planet B," but talk is cheap. Do these protesters practice the basics? How much did they spend on things that fund extractive industries and their lobbyists?

Corporations that pollute and deplete like Exxon and Apple don't buy their own products. Consumers do. *We* do. CEOs and politicians

---

11   Alejandro de la Garza, "We Have the Technology to Solve Climate Change. What We Need Is Political Will," *TIME*, April 7, 2022.

know that someone who says one thing but does another will buy and vote with their behavior over their words. How can we blame executives and politicians for responding to our collective actions?

It Looks Hard: The Stages of Stewardship

People respond to our environmental problems in stages. At first, many don't see problems. Some then see them. Some who see them act, and so on in stages of what they do, fewer at each stage. A few reach the stage of seeing that even if we could restore all our environmental problems to pre-industrial, if we don't change our culture, we will reproduce our current problems. That is, any solution must change culture.

Then nearly everyone at that stage concludes that changing culture is too hard. They revert to an earlier stage and decide to work on something they consider more achievable, like technology, efficiency, legislation, or something inside our culture.

STAGES OF STEWARDSHIP

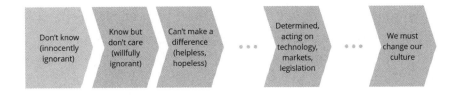

Making elements of our culture more efficient won't change it. On the contrary, if we make a polluting system more efficient, we will pollute more efficiently, what I call *stepping on the gas, thinking it's the brake, wanting congratulations*. Like driving with a flat tire, working *in* the system instead of *on* it, believing, "I may not be doing as much as you, but at least I'm doing something" leads to working hard and going in circles.

Can we change culture? Yes, people have changed cultures deliberately at the level of nation and empire, including abolitionism, ending Apartheid, and establishing LGBTQ+ rights, however incomplete so far.

I traversed these stages too. I grew up in the "innocently ignorant" stage. My parents showed little interest in sustainability beyond putting the ironic-at-best "Boycott Exxon" sign in the car's back window after the Valdez spill. We just bought gas from Exxon's competitors. I reached "can't make a difference," so felt helpless and hopeless for three decades. I stopped eating meat in 1990 partly for environmental reasons, but, again, faced friction from my family and culture. Meanwhile, I flew around the world and bought what I wanted when I felt like it without regard to its environmental impact because what else was progress for?

Chapter 3 recounts my reaching later stages. It still took years swimming upstream to do more than step on the gas thinking it was the brake. I took several more years to see changing culture was possible and desirable, then several more to develop strategies and tactics that worked, then more to find ways to teach and train it. Now people feel gratitude and want to teach and train others—essential for a movement to power itself. Now transformations that took me years, people do in weeks with the Spodek Method.

Still, changing culture, even if possible, seems like a lot of work. Meanwhile, going with the flow looks more comfortable. People stay at earlier stages because later stages look like struggle.

As American abolitionist, social reformer, and once-enslaved Frederick Douglass said,

> If there is no struggle, there is no progress. Those who profess to favor freedom, and yet depreciate agitation, are men who want crops without plowing up the ground. They want rain without thunder and lightning. They want the ocean without the awful roar of its many waters. This struggle may be a moral one; or it may be a physical one; or it may be both moral and physical; but it must be a struggle.

Progress and freedom sound nice, but with all that struggle, the choice to act or not looks like:

INTEGRITY UNDER ADVERSITY, SHORT-TERM

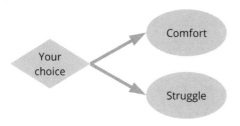

As it turns out, seeing the choice as between comfort and struggle is only the short-term view. To struggle may start hard but leads to freedom, increased self-awareness, and love. Comfort in a culture that violates our values (and its own) may bring short-term convenience and acceptance but it leads to giving up, giving in, lower self-awareness, and ultimately self-loathing, suppression, and denial. In our culture it also leads to pollution, misery, and sickness. By contrast, struggling to live by your values in a culture of giving up forces you to understand why you struggle, which creates self-awareness. In a polluted world, living by the values of stewardship creates love and health. Achieving your goals against adversity creates glory.

It Gets Better

Longer-term outcomes are the opposite of short-term perception here. The earlier choice that appeared to be between comfort and struggle was actually between *perceived* comfort and *perceived* struggle:

INTEGRITY UNDER ADVERSITY, LONG-TERM

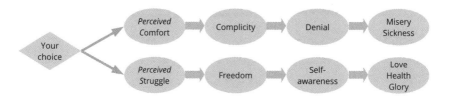

Any parent who has had to wake up at four a.m. to change their baby's diaper knows this choice. When the baby cries, it's more comfortable to stay in bed, but the effort to get up and change the diaper is more than worth it. Dog owners who leave a party early to walk the dog know it too.

It Gets Even Better

I value struggle and have benefited from it, but I've found the struggle in acting sustainably decreasing the more I practice. After the mindset shift of the Spodek Method leads us to find intrinsic, in-the-moment reward, we enjoy our efforts. What once felt like struggle becomes engaging, even fun. We no longer feel comfortable violating our values. We identify that comfort that leads to giving up with procrastination.

After the mindset shift, the choice becomes between procrastinating and engaging in purposeful effort:

INTEGRITY, POST-MINDSET SHIFT

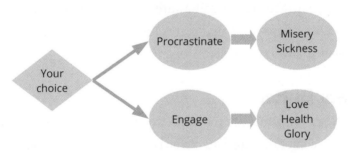

Is it suffering to leave a party to take care of your baby or walk your dog as an act of love?

CHAPTER 2

PURPOSEFUL EFFORT
AND MEANING

In a culture of denial and resignation, we need to choose to stop procrastinating for short-term convenience and engage. We can only change that culture by changing ourselves as individuals first. It's one thing to say that purposeful effort creates meaning abstractly, but another to live it, which can be grueling. If you've overcome adversity, you know what I mean. I'll share examples from my life that opened me to see great examples from the history of people who procrastinated and who have engaged to live by their values.

Surviving Despair

The booming economy of the 1990s led me from interesting but academic research to invent a technology that led to co-founding my first company, Submedia. We created jobs and satisfied customers, reaching a thirty-million-dollar valuation (on paper). I learned to value trade, technology, and innovation.

I conceived the idea in graduate school. At the time I was helping build a satellite with NASA and the European Space Agency. Two friends approached me to try our hands at entrepreneurship. I had the idea for an optics technology that would show motion pictures to people moving past it. We installed displays using it on subway tunnel walls to show movies to riders between stations. The market potential was huge: subway systems own real estate in cities, transporting large audiences that advertisers value because riders in trains can't escape being right in front of their messages. We envisioned selling ads and sharing revenue that could offset rider fares. I felt some misgivings about putting ads in yet another space, but nearly no one else did. On the contrary, rider surveys showed nearly all respondents found the ads beautiful and preferred them to dark walls. They liked that ad revenue would offset their fares too.

We filed our first patent in 1998. Our first investment came in 1999. Atlanta's subway system signed in 2000. Coca-Cola signed in 2001 as our debut advertiser, beating Nike by an hour.

I started living like a successful businessman, spending on higher-end clothes and fancier restaurants. I flew to camp in Yosemite and party in Miami. I dreamed that when Submedia cashed out, I'd fly my friends to celebrate overseas.

We hired a major New York PR firm to elevate the Atlanta launch into a global media event. Coca-Cola was fully on board. They wanted to launch in Atlanta's summer, but a Tuesday after Labor Day meant the most chance for media coverage.

By launch, the *Wall Street Journal* had covered us twice, including a four-column spread. The *New York Times* and live television coverage looked likely. We mailed postcards to invite stakeholders. Here's the front, featuring a frame from the ad of Dasani water flowing:

Here's the other side. Note our launch date:

> **SUBMEDIA**
>
> Partners and friends, please join us for
>
> **THE U.S. DEBUT OF**
> **IN-TUNNEL**
> **MOTION PICTURE**
> **ADVERTISING**
>
> In conjunction with Coca-Cola and the
> Metropolitan Atlanta Rapid Transit Authority,
> Submedia cordially invites you to the historic
> launch of our revolutionary medium in Atlanta, GA.
>
> 10:30 am, September 11, 2001 at the
> Concourse Level of M.A.R.T.A.'s Dunwoody Station
>
> For directions, hotel information, and
> other details, email info@sub-media.com.

Yes, we had planned to launch September 11, 2001. Needless to say, we canceled.

I had spent the night before in the tunnels making sure everything worked and was safe. Everything checked out. We had prepared for all we could. Since I owned about a third of the company then, my net worth at eight a.m. was about ten million dollars, on paper. By noon it was zero.

Submedia eventually launched, but a subdued event in October. Subway systems denied access to build for security reasons. The tech bubble bursting, a recession, the advertising market collapsing, and investment drying up put us near bankruptcy. I had made mistakes as CEO, as did the team, but the technology worked and we delivered what we promised to advertisers and subway systems. We had grown our staff fast, anticipating growth, so we had to lay off people who had done great work. I shed tears, feeling guilty. Some who remained worked to salvage what we could, but others split into factions, trying to get what they could. Board members screamed at each other. I didn't eat enough; my well-being seemed unimportant.

As a first-time CEO, I had confused authority with leadership. I didn't know how to unite and motivate under adversity. Soon enough, the investors squeezed me out. I felt (and still feel) shame and guilt for being unable to repay shareholders who invested in a dream I sold them so I kept coming to the office to try to help. When the new leadership moved to a smaller office, they simply didn't give me a key. I felt like a dog they kicked that kept coming back.

I had lost all I cared about. I had started graduate school dreaming of following Galileo and Einstein in discovering new phenomena or laws of nature. To start the venture, I had abandoned those dreams. After losing ownership of my invention and company, all I had were a few connections in outdoor advertising. No longer focusing on building my company, I came to see the field as often blocking views of nature, promoting sugar water to kids. I had allowed myself to overlook these aspects of outdoor advertising, anticipating wealth and comfort. That is, I allowed myself to be corrupted.

But how could I compare my loss of mere dreams and money on paper to what so many lost on 9/11? I couldn't, so I didn't complain. Across the street from my home is a firehouse with a plaque commemorating the firefighters who on 9/11 entered the burning towers and never came back out. One of my oldest friends, later the co-founder of my second company, served in the first Gulf War as a marine. He had enlisted to defend his country and freedom, as did many after 9/11. All of these people had families.

I didn't allow myself to grieve. Only decades later, when I started sharing this story, did friends point out that because others lost more didn't mean I couldn't grieve my losses. For years I saw little to live for, nor any foothold to rebuild from or anyone to share my despair with.

Despite the hopelessness and despair, I had to eat and pay my mortgage. I had to make money soon. I had accepted only enough salary from Submedia to live on, keeping the rest in the company, which investors had squeezed me out of, so I had no savings. In one sense it helped to have one course of action: get a job, pay my bills, create a new future, and stop clinging to the old one.

Submedia after 9/11 was not the greatest challenge I lived through. One came in the mid-to-late 1970s when I was in grade school. My parents struggled financially before their divorce, then more afterward. They agreed on joint custody of me and my two sisters. My mom was briefly homeless, though she didn't describe it that way then because she kept a roof over our heads by bouncing between friends' spare rooms. Eventually she bought a house on Rockland Street for five thousand dollars. The house was that cheap because it was in what novelist Mat Johnson 🎙, who grew up a block away, called the "ghetto" and "the heart of black Philly." Her mother, my grandmother, wrote in her journal after her first visit: "I silently fumed, 'How could she invite me to spend my vacation in these slum conditions?'"

I visited Rockland Street while writing this book. People live in the house we used to live in, so I won't share pictures of it, but the picture on the left below shows a building a few doors away whose boarded-up and shattered windows looked similar to when we lived there. The picture on the right is a house my sisters and I could see from our school bus stop. A current resident said, "Rockland Street is a real shithole right now . . . real crime and mayhem. At a certain point we got bullets in our home through the foyer [and] front door."

As far as I know, everyone on Rockland Street was poor, but I should point out that while my mom lived there and in her next home, on Walnut Lane, my father lived in a middle-class neighborhood. He had a secure if modest income as a tenured professor at Temple University. But on Rockland Street, in the summer, the city provided welfare sandwiches at one end. The other end, neighbors warned my mom to keep me away from, where the gangs were. Broken glass was everywhere. She had to work so we were latchkey kids, catching the bus ourselves in the morning and taking care of ourselves after school. Once, my mom found out that we came home hungry and why: other kids stole our lunches while we waited for the bus. The girls didn't let my sisters play Double Dutch with them, saying white girls didn't know how. Girls would touch my skin, remarking on how light it was.

Only decades later did I learn that most people weren't mugged once while growing up, let alone many times. I was. They didn't know what being threatened by gangs brandishing wrenches and rocks in their face felt like. They haven't been pick-pocketed, then sucker-punched in the jaw while waiting for a bus. I didn't know that most people didn't feel the shame of feeling "I should have fought back" then being blamed or teased for being a victim. It never occurred to me that most white people hadn't spent time as racial minorities targeted and victimized with impunity. They didn't know life without recourse to justice as I did in for those years.

To clarify, I've since lived most of my life in the majority, and recognize the advantages I've had in other situations. Still, here is my junior high school class (photo on right). Today, when I ask people who the minorities are in the picture, they generally say the non-white students. As one of two white boys out of a couple dozen students, being part of a small minority was normal for me. My high school had fewer white students than black students too. Once, walking to the subway from school with the other white boy, who was eating a snack at the time, another boy grabbed the snack, threatened him, and took it. He was mugged right in front of me. To this day, I feel shame not to have helped him.

I don't want to imply the situation was all violence and exclusion. My mom loves connecting with people and got along with our neighbors. She had taught elementary school on the impoverished South Side of Chicago in the turbulent 1960s. Her students called her black because, as other teachers told her, no other white people treated them respectfully. In the farm and town she grew up in in South Dakota, she had several Native American friends. When, as a teacher, she took the Chicago kids on a field trip to the South Dakota reservation her friends grew up on, they said, "We thought we had it bad."

My mom remarried a man who also wanted to make Rockland Street work. While growing up outside Philadelphia, his neighbors burned a cross on their lawn to threaten his family for working on civil rights. On Rockland Street, my mom and stepfather helped organize neighborhood activities. Neighbors took us to gospel church with them. I remember stickball and basketball played with a milk carton bolted with a plywood backboard to a telephone pole. I loved the welfare peanut butter at neighbors' houses so sweetened, it tasted like candy.

I considered my experiences "real," teaching me what life was *really* like. It helped me understand what Nelson Mandela meant by "black domination" in the speech he said to the judge before being sentenced to twenty-seven years in prison:

> During my lifetime, I have dedicated myself to this struggle of the
> African people. I have fought against white domination, and I have
> fought against black domination. I have cherished the ideal of a

democratic and free society in which all persons live together in harmony and with equal opportunities. It is an ideal which I hope to live for and to achieve. But if needs be, it is an ideal for which I am prepared to die.

As we'll see in part 3, dominance hierarchies play a major role in our environmental problems and the underlying loss of freedom, reason, and democracy. Seeing in my formative years that people of any skin color could dominate or be subjugated led me to wonder what led people to do it. Finding that answer uncovered solutions. I believe this curiosity helped drive me to find it.

Still, when some boys stuck a lit firecracker that exploded in my pocket, my mom and stepfather decided they couldn't make Rockland Street work and moved. The new home, about a mile away, on Walnut Lane, was safer, but still on a block where my mom had been mugged at knifepoint before we moved there. I would be mugged there too, as well as by my father's house.

My mom has pointed out that legacy of redlining—the real estate practice of excluding non-whites from access to homes and home loans in some neighborhoods—meant we could get out, so Rockland Street had been less an inescapable ghetto for us. Still, from my perspective, those who excluded me from their social groups had higher status. In that world, being white and a boy meant being victimized with impunity, with the intersection yet more acute.

Speaking of ghettos, my father's parents came to America from Poland but he clarified that because they were Jewish, they weren't Polish citizens. They likely lived in ghettos. I doubt my Rockland Street experience was as confining as theirs. Neighbors up the block from my father's house had escaped Hitler, as did a distant cousin I met, Zipporah. I presume the rest of the family they left behind ended up in Auschwitz.

A specific event from that time helped me recover from Submedia's collapse. When she was staying with a friend before Rockland Street, I saw my mom cry. I was too young to know it at the time, but she couldn't afford to make ends meet. I had seen her cry like that before. Referring to that time, I asked her, "Is it as bad as the time the refrigerator broke?" It was an innocent question, but I happened to refer to her rock bottom since the divorce. It prompted her to think that since she had survived that time she could survive this one too. She stopped crying and hugged me.

Back to post-9/11, eventually I found a job. In time I learned to use my experience to persevere and create meaning too. I never felt with Submedia that I faced the adversity my mom or my father's family did. We've all struggled. You who are reading these words have struggled too. Restoring sustainability to our culture will take overcoming procrastination followed by engaging. Helping each other using what we've learned helps us persevere and create meaning.

Engaging Brings Meaning. Procrastination Doesn't.

You may have noticed that I lamented losing my dreams to follow Galileo and Einstein when Submedia collapsed, not when I left physics. Why then and not before? Likewise, for most of my life I knew about our environmental problems but remained passive about them. So why did I work hard to put my life back together after recovering from Submedia's collapse after 9/11 but not on our environmental problems for my life's first four decades?

I learned the answer from various sources, but one of the most important was from a conversation in 2021 on my podcast with Blake Haxton 🎤. I learned from his TEDx talk before meeting that in high school, he had a persistent pain in his leg. He went to a doctor. The doctor ordered him: *get to a hospital now, as fast as you can!* He did. Six weeks later he woke up from a coma without his legs. Doctors amputated them to save him from flesh-eating disease. He had been minutes from dying.

His loss seemed greater than mine with Submedia. Did he feel sorry for himself? On the contrary, he had just won a silver medal in the Paralympics. He described himself as lucky.

"Lucky?" I asked. "You lost both your legs."

"Yes, but that was just one thing."

"It was a big thing."

He explained that he learned that everyone has to struggle. Many people's struggles never get recognized or others treat them as meaningless, but everyone sees how much he's had to overcome. Some struggles give life meaning. It's easy to presume we've worked harder than others or they've had it easier. I realized why I felt so comfortable with him. He took for granted that I struggled, which made me feel appreciated, understood, and supported.

Blake shared life lessons, as I saw it, on the level of Viktor Frankl, whose book *Man's Search for Meaning* described his surviving Auschwitz by creating meaning. Both Haxton and Frankl were human, with no special ability to create meaning from struggle. Blake's legs didn't define him, nor did Submedia's success define me. The problem wasn't losing those things but clinging to a world no longer viable. Likewise with the environment and our culture. We don't need to lose our legs or survive Auschwitz to create meaning from struggle.

I've learned not to value my life based on things outside of my control, but on how much I reach my potential. We can all grow and create as much meaning, purpose, and satisfaction as we want if we don't cling to a future no longer viable. In this book, I'll show that if you choose to stop procrastinating in order to live sustainably, you'll help alleviate the suffering of billions.

By the time I spoke to Blake, I had reduced my environmental impact by over 90 percent and learned sustainability didn't mean deprivation, sacrifice, burden, or chore. I struggled at first, but found meaning and stopped procrastinating. Then I found freedom, joy, fun, saving money and time, and more. Blake helped me realize why I am living a better life today. I gained living by my values, liberation, and freedom. As one example, since my diet moved to fresh local produce and my taste buds recovered from my old diet, apples taste sweeter than ice cream did. Vegetables taste sweeter than fruit used to. Less sugar and more sweetness seem too good to be true (likewise less salt but more salt flavor, less oil but more pleasure, more volume of food but less fattening), but it keeps happening in many ways beyond food.

It didn't take long, maybe a couple months into my experiment of avoiding flying, before each step made the next appealing. I didn't need willpower to keep going. At a certain point it would take effort *to stop* because of how much I was learning about cultures, nature, history, anthropology, and so on. I felt like I had mastered playing an instrument and wanted to explore new songs and genres.

Maybe you prefer to procrastinate. You might point out that for Blake, the flesh-eating bacteria forced his change. You'll act if a disaster forces you to. You might feel as I did about the environment: "I didn't cause the problems. Past generations did. Others are polluting and depleting more than I am. *They* should change. It's not fair that others are benefiting while polluting and depleting more while others are suffering more while polluting and depleting less."

These thoughts and more discouraged me. Considering them today, I recognize how they comforted me then, but now they sicken me in how they promote giving up. Nobody without hands-on experience believes me on this view, but post-Spodek Method, we love talking about such transitions and finding new ones. Having found meaning from engaging and having seen others do it too helped me prepare to stop procrastinating and engage in sustainability. I began to find role models from history who revealed what we're capable of.

Living By Our Values Enables Us to Lead

People sometimes cry in my workshops when they realize they can be more than great proponents of freedom like Thomas Jefferson or James Madison. They can combine words like theirs with the action of Harriet Tubman and the integrity of Robert Carter III.

Thomas Jefferson was one of the great writers on freedom and equality. When I first heard of people taking down his statue, I thought they were overreacting. Then I read his racism in his *Notes on the State of Virginia*, suggesting Africans mated with orangutans, for example. If he loved freedom so much, how did he not free his slaves? As a president from Virginia, imagine if he had. Had he freed his slaves, might he have been able to influence George Washington and James Madison to free theirs? Might they have influenced the Constitution to have helped ban slavery without the Civil War?

Instead, Jefferson didn't practice freedom and equality. He could have. Washington freed his slaves in his will. Benjamin Franklin had owned slaves, but by 1750 spoke against slavery and in 1774 co-founded an anti-slavery society.

Harriet Tubman was born into slavery. After escaping, she could have laid low, which would have been safer. Instead, she returned to slave territory to free other slaves—thirteen missions saving about seventy people. You could say that out of four million slaves, seventy didn't matter much or what people say today: "individual actions don't matter," she "wasn't changing the system," and so on.

Do we believe ourselves when we say individual action doesn't matter? Would you put your life on the line for that belief? I know my actions matter, which is why I do what people call extreme, but I haven't staked my life for it. She knew her actions mattered enough that she did stake her life on it. How did she know? For one thing, she knew she wasn't alone. She worked with the Underground Railroad, which saved about 100,000 people. You might say that even that number is small compared to the total slave population of four million, but the Underground Railroad was part of a global abolitionism movement. Their work contributed to electing a president who came to see the need for a constitutional amendment and did what it took for it to pass. Tubman's missions enabled her to serve in the Civil War as a spy, scout, and as the first American woman to lead armed expeditions.

Unlike Tubman, we don't have to risk our lives to act in the field of sustainability. On the contrary, it improves our lives and communities.

Still, when we feel like giving up, when people tell us what we do doesn't matter, her words can inspire us:

> If you hear the dogs, keep going. If you see the torches in the woods, keep going. If there's shouting after you, keep going. Don't ever stop. Keep going. If you want a taste of freedom, keep going.

But we today benefit from a system that hurts others. Tubman didn't own slaves. A more relevant role model may be Robert Carter III, though you may not know his name. In 1791, he began freeing his five hundred slaves, the most of any American, including giving them land. He didn't do it for a marketing campaign, to make his quarterly numbers, or to follow a trend. The decision took reflection, conversations with family and friends, as well as gut checks. His neighbors resented him for it. He was rich, so he wasn't risking bankruptcy, nor was he a saint, but he did what he believed was right.

He is a relevant role model for us who pollute and deplete. My point is not to compare abolitionism with sustainability but with our place in the system. Carter was a classmate of Jefferson and knew Madison and Washington. Imagine if he had influenced them. Might he have helped influence the Constitution? What influence might we have today? Even if we believe we can't influence anyone, don't we want to live by our values, with integrity, as Carter did? His being born into status enabled him to influence at points of great leverage. We have access to points of leverage too.

We'll learn of other role models who were members of the oppressing group who could have remained complicit but instead worked against the system their peers accepted. They changed nations and empires on the time scale scientists say we have to act in. We didn't create our culture of pollution and depletion that undermines freedom, democracy, human ingenuity, health, and security, but it exists and has corrupted us as much as slavery corrupted Jefferson, Madison, and Washington. We can procrastinate to perceived comfort or we can engage to create meaning, value, and purpose. When we do, we will feel what Tubman did when she entered free territory:

> When I found I had crossed that line, I looked at my hands to see if I was the same person. There was such a glory over everything; the sun came like gold through the trees, and over the fields, and I felt like I was in Heaven.

CHAPTER 3

GROWING UP IN OUR CULTURE

When I tell people my parents helped start a food co-op during my childhood in Philadelphia, many conclude I was raised in a family active in environmental conservation. I wasn't. They started it because they didn't have time or money and had to feed three kids. I'm still surprised when people today associate co-ops with costing more time or money because I experienced it as the opposite.

My parents met in India, both on government grants, my father to study history, my mother to teach English. My older sister was born there and I spent a year of my childhood there. People then flew less than today and I felt special at having grown up flying as well as entitled to fly more. I came to see it as good, believing it helped me learn about other cultures and for them to learn about us.

My father polluted plenty flying around the world. Despite his modest income as a history professor, he was good at getting grants to research abroad, sometimes bringing us. After retiring, he had more money and flew around even more.

My mom and stepfather struggled, but when their careers enabled them to afford it, they polluted plenty. They recently bought a luxury SUV and flew to the Caribbean when a friend who lives within driving distance held a birthday party there. Growing up, though, we only got our first air conditioner when I was around ten. We could afford to turn it on only a few nights a year and on those nights the whole family slept in that room. We didn't have cable or color TV. Everyone took turns cooking. We kids loved McDonald's, but we rarely went—not to be sustainable. We just couldn't afford it.

My parents wouldn't buy sugar cereals or soda, but I loved them. I'd gorge on them when I visited friends. I got chubby to where my stepbrother would pinch the fat under my nipples and twist, saying "Titty twister!" I felt shame, though when I grew up I still always had ice cream in the freezer and chips in my cupboard into my forties. I ran marathons to try to lose the fat around my middle, but couldn't outrun my diet. I created elaborate rules to limit myself. Instead, I felt more shame each time I finished another container of Cherry Garcia.

Growing up, I had a friend who lived a short walk from my father's house. His family lived in a big house with central air conditioning. Their Volvo was nicer than my dad's Toyota. They bought him an Apple IIe and an SLR camera. I envied him. I learned from him: "whoever dies with the most toys wins." I wanted more toys.

I grew to have proudly flown to six continents and over two dozen countries including North Korea twice. My family had spread all over the world, as had the families of everyone I knew. My income depended on flying. I knew my flying polluted but figured it wasn't that much, especially compared to the cultural exchange it fostered, which I saw as good. I saw not flying as holding back progress. I felt we should help everyone to fly.

Submedia's early success starting making me feel I could fulfill my dreams of having more toys. Discovering nicer clothes and furniture led me to see how much more there was to "more": bigger apartments, higher-end furniture, and fancier clothes. My dream to fly my friends overseas was more to impress them than to enjoy myself.

I didn't see a problem with wanting more. On the contrary, most of my life I wanted even more. While finishing my PhD and starting Submedia, I audited classes in entrepreneurship at the business school. There I learned about free trade. If I produced something you valued and we agreed on a price where I valued your money more and you valued my product more, we both benefited from trade. My profit meant I benefited the world. More trade and profit meant more good.

Did I worry about the environment? I knew for as long as I can remember about global warming, acid rain, smog, plastic pollution, and our other environmental problems, but I lived in Manhattan without a car so I figured I impacted less than most.

I didn't act, though, and thought of many reasons not to. Some thoughts: "To reduce my impact would deprive me significantly while barely affecting the world." "It makes no sense for one person to suffer for negligible global change." "Meaningful effective results require governments and corporations to change. *They* gain power and profit from polluting and depleting. If anyone should change, *they* should, not me."

Other examples: "What if we Americans acted more sustainably but competing nations didn't? Wouldn't they make up the difference? What if they beat us economically or even militarily?" Moreover, "Businesses have incentive to become more efficient. Can't I expect market forces to drive down waste?"

Or: "Markets allocate resources to people that solve problems best. Markets help lift people out of poverty." By my adulthood, even poor people had color TVs that we couldn't afford growing up. I grew up with a rotary phone. Not everyone lives near a farmers market. Amazon.com may overuse packaging, but doesn't it save people money? Do rich people understand the challenges of people who have to work? I was doing all I could. I put my faith in others to solve our problems.

School Barely Taught Sustainability

My earliest memories of sustainability come from my fifth-grade teacher, Mrs. Fingerhood. My sisters and I in turn passed through her classroom. She taught us generally about nature's beauty and humanity's impact, and bird-watching specifically, but not much about action. At that time, my education contained little on sustainability.

I didn't pursue graduate school in science for sustainability. I loved the beauty of physics and wanted to discover things in nature as Einstein and Galileo did. I didn't know then that my science training would apply to sustainability.

I stopped eating meat only partly for environmental reasons. It also saved money.

I didn't pursue an MBA to learn leadership either. I did it to get rich. I didn't know classes in leadership existed until business school. I didn't anticipate learning leadership would apply to solving humanity's biggest problems.

In other words, I was lucky that my life choices gave me experience in domains critical to lead effectively in sustainability, though just having that experience led me to act. Only when my leadership coaching led to clients in the c-suites of publicly-traded companies did I discover the value of integrity and personal responsibility to leadership—that I can't lead others to live by values I live the opposite of and that systemic change begins with personal change. Seeing the integrity and personal responsibility that brought my clients to the tops of their fields leading organizations inspired me to require more of the qualities in myself. Then my procrastination on sustainability began to nag at me.

Then one day in the fall of 2014, I looked at my garbage bag in my kitchen and thought, "That's a lot of garbage." I filled about a load a week then, which felt normal. I noticed most of it was packaging and wondered if I could produce less. Even if I couldn't solve the world's trash problems myself, could I take responsibility for my contribution?

I wondered if I could go a week without packaged food. I figured I could in principle. Could I do it in practice?

My First Experiment: Avoiding Packaged Food (2014)

As latchkey kids, my siblings and I were assigned to make dinner for when the grown-ups got home, so I grew up cooking—just not from scratch.

Cooking generally meant combining the contents of cans, boxes, bags, or jars with some fresh produce. As an adult, I used less fresh produce. I bought a lot of cans of chili, plastic containers of hummus, and takeout.

The challenge felt like a possible learning experience, not that I saw much point since my views then told me that I would sacrifice a lot personally for negligible effect on the world. Only governments and corporations could change on the scale necessary to make a difference. People train their whole lives to become chefs in Manhattan. Foods are imported here from around the world. Why not enjoy them? Isn't enjoying life what progress was for?

Still, what was a week? Eventually I got to logistics. How exactly would I do it? I started trying to plan each day and what qualified as packaging. Should I count stickers and rubber bands?

Each time I almost started, I'd doubt myself. Analyzing and planning to answer these questions made me feel productive, but I didn't act. For *six months* I analyzed and planned without acting until one day I thought, "I'm not going to die if I eat only fruits and vegetables for a week. I'm starting now." And I did. Instead of solving all the logistics before starting, I figured them out as they arose. For example, I decided not to buy packaged food, but could finish packaged food in my cupboard. I disallowed stickers and rubber bands. I allowed bulk foods in reusable bags I brought.

My first gut check came midweek, when I ran out of food at home and went to the store. For the first time, I noticed I saw almost no food on the shelves. I saw boxes, cans, bottles, bags, and jars, but not food itself. It hit me: I may have never eaten a meal without something packaged— meaning without polluting. For all my degrees and having helped build a satellite, a thirty-million-dollar company, and so on, I couldn't eat without hurting people. I couldn't *live* without hurting people.

I stood speechless. My mind struggled to try to make sense of this . . . this what? Contradiction? Predicament? Challenge?

I would have stood longer, but I was hungry. I looked at the produce section and realized all I could do was buy vegetables and figure it out.

Then I realized I could buy dried beans from the bulk section, where I had only bought dried fruit and nuts before. I'm not proud that it took me into my forties to do it, but that week I soaked and boiled beans for the first time. I served them with steamed broccoli. It wasn't glamorous, but instead of analyzing and planning, I solved the actual problem I faced. I felt foolish for taking six months to start.

I made it two and a half weeks before buying my next packaged food—a bag of onions I could have bought loose. Not long after, I bought a can of tomatoes. I realized I would have preferred fresh and it became the last can I ever bought. I decided to keep going, not at zero, but at my best efforts. For months, my food was bland: mainly steamed vegetables over lentils or beans with salt, pepper, and maybe vinegar, sometimes awful combinations. At first, shopping and cooking took longer. I experimented and started learning what flavors and textures I liked. At my sister's suggestion, I joined a subscription where each week I picked up a load of produce from a local farm (called a Community Supported Agriculture or CSA program). Accepting what the farm sent forced variety on me. I also started shopping at farmers markets, which I had seen as snooty and expensive before. At first I couldn't tell kale from collards and felt scared to try new vegetables. In time I learned local produce and enjoyed trying new things, even weird ones like stinging nettles. I've also learned to save money by shopping seasonally, meeting the farmers, and getting discounts for imperfect produce.

I also learned to cook. I found out that my rice cooker cooked beans too. It still took as long as boiling, but I didn't have to watch it so my preparation time dropped to a few minutes. Later I bought a pressure cooker. It cooked beans in minutes, which changed everything. Cooking from scratch became cheaper, faster, and more convenient than restaurants or takeout. I discovered that *cooking* doesn't take longer or cost more, *not knowing how to cook* does. I can't believe how backward I'd gotten it. When I was ignorant and bought prepared food I saw cooking from scratch as expensive, time consuming, and privileged. Actual practice showed I had it exactly backward. (Years later I saw why, as I'll cover in part 3).

More importantly, it started tasting better too. I developed what I called my famous no-packaging vegan stews. After six months, I loved cooking from scratch. I invited people over, though early on I was still inconsistent. Within a year, I had found consistency.

I went from emptying my garbage weekly to biweekly, as I learned more ways to avoid garbage, then to monthly, to yearly, and kept going. I saw Lauren Singer's TEDx talk where she fit a year of garbage in a jar.[12] At first I thought it was a trick. Then I wondered how she did it. What if it wasn't a trick? What if my thinking it was a trick was my mind defending itself from facing that I was acting against my values and that my claims of helplessness were false? Then I saw her as a role model. Typing these words, I'm in my fifth year on one load of garbage.

When I look back after years of enjoying fresher, cheaper more delicious unpackaged food, I see that deliberate action changed my identity. I realize I had subconsciously wanted my experiment to fail. I wanted to say I found the cure worse than the disease so I could stick with the disease. I wanted to justify procrastinating. I didn't expect engaging to bring meaning or deliciousness, saving time, or saving money, but it opened my mind to receive it.

Avoiding Flying (2016)

In January 2016, on a flight home from a vacation, I happened to watch a video of a presentation on sustainability.[13] The speaker was more than credible: Caltech-trained Cambridge University physicist David MacKay. He said that flying round-trip from London to LA impacted the climate as much as a year of driving. The comparison depended on many variables, like what car someone drove and how often, but he was speaking generally.

---

12 "Why I Live a Zero Waste Life," TEDx Talks, May 27, 2015.

13 "Sustainable Energy—Without the Hot Air with David MacKay," Harvard University, October 25, 2010.

The comparison was a gut check for me. I thought living in Manhattan without a car meant I impacted less than most Americans. Instead, I was impacting as much, with my greatest contribution, flying, coming from something I thought was good. I had found this information by chance, but I could have looked it up earlier. I procrastinated. I felt guilt and shame at violating my values and keeping myself willfully ignorant. I had to ask myself: Who was I? Was I someone who looked the other way to avoid facing the impacts of my actions on others?

As with avoiding packaging, when I thought about avoiding flying, my mind filled with questions about my internal conflict. I subconsciously crowded them out with analyzing and planning but not acting until my next flight, to give a presentation in London and visit Paris. I had lived in Paris from 1990 to 1991 and had looked forward to visiting my former home for the first time since. Instead I couldn't sleep. I wondered if an experiment like avoiding packaged food for a week could work here too. Could I avoid flying for a time? A week or month seemed too short to be meaningful. Could I avoid flying for a year?

A year?!? The thought nearly stopped my heart. How could I avoid flying for a year? Why should I, anyway? What about family who lived flying-distance away? How much income would I lose? How could I be a good citizen? What about adventure, cuisine, and fun? Would it make a difference?

But I couldn't help wonder if my unexpected positive results in avoiding packaged food might appear here too. I couldn't imagine it, but I also couldn't sleep because I knew what choice I considered right.

I returned from Paris March 22, 2016. I committed to avoid flying for one year, starting March 23. As with avoiding packaged food, I knew I wouldn't die. I could always quit and fly if I had to. But how would I live? Was an experiment worth going broke and my family disowning me?

Within three months, to my surprise, the food experiment results repeated. I didn't spend the time not flying staring at the wall or lamenting. I came up with new things to do with my newfound time and money. To my amazement, I found myself connecting *more* with family. I found more adventure and cuisine without flying.

What was going on?!? I couldn't believe the results except that they fit the pattern from the first experiment. My mindset shifted. I came to see earth and our population as too big and beautiful to see everything and everyone, even if I flew every day. Even if I did, any time I visited one place or person, I wasn't visiting another. Flying didn't enable me to see everyone and everything. It just made me think I could—wrongly. Not flying didn't limit me. Life did. Choosing to avoid flying led me to choose what and whom I would see deliberately. I found that my best travel strategy was to maximize my joy where I was and with whom I was at all times, which led me to enjoy everything more, even in one place.

It took years of not flying, but I saw my old attitudes about flying as craving and anxiety, based on a fear of fearing missing out. For years I would think of what I was missing and doubts would resurface, but I increasingly saw myself living in abundance regarding *travel*, which I distinguished from *flying*. I took sailing lessons to get off North America (before Greta Thunberg sailed across the Atlantic). Before then, I had thought only rich people sailed. Again, my ignorance led me to see an alternative to flying as privileged when hands-on experience showed I had it backward. Flying doesn't help spread culture, it homogenizes it. It doesn't help the poor, it makes people poor. I came to see the feelings of entitlement and self-righteousness with flying. I discovered in sailing an activity thousands of years old practiced everywhere. Just stepping on a boat was like a vacation, more accessible, cheaper, and cleaner than flying. I saw the travel industry pushing dissatisfaction of where I was and whom I was with. Flying and the culture around it started feeling wretched, even disgusting. I took the train across this beautiful country. I biked. I camped. I found diverse cultures and nature within commuter rail and biking distance.

Three months in, I decided to extend the experiment to another year. The next year I extended another and so on until I decided never to fly again, no willpower needed. I just kept finding more liberation and followed the path to more freedom.

Avoiding flying led to decreasing my footprint over 90 percent in one two-and-a-half-year period, measured by an online calculator.[14] Contrary to my expectations, reducing my impact *improved* my quality of life. Engaging instead of procrastinating was beginning to create meaning. (On terminology: the Global Footprint Network, whose online calculator I used, uses the term *footprint* to describe all of one's effects on the environment, not just carbon, greenhouse gases, or global warming. I use the term *impact* to describe my overall impact, not just warming. I use the term *footprint* here to keep it associated with the calculator.)

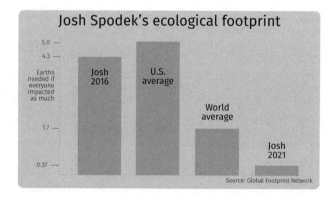

I always have to clarify that this graph illustrates to me freedom, fun, joy, meaning, and purpose, since most people I show it to presume I deprived myself and sacrificed. I didn't. I thought I would—an expectation I now see based in my ignorance and inexperience. I'll show in part 3 why my beliefs predicted precisely the opposite of what happened. People also wonder if it was a trick, as I thought of Singer's TEDx talk, or if I had special advantage. Since I was already vegan, didn't own a car, and didn't have children to buy toys, diapers, or clothing for, I couldn't reduce the

14 "Ecological Footprint Calculator," Global Footprint Network.

biggest ways available to most people. If you eat factory-farmed meat, drive, and have kids, you can likely reduce more, faster, and enjoy more improvement to your quality of life. If it sounds impossible, it would have to me too.

Picking Up Litter (2017)

Litter is drowning our world. At least litter in the past wasn't plastic, which doesn't biodegrade but causes birth defects and cancer, kills wildlife, and more.[15] I wished I could help reduce it, but after a couple years avoiding packaged food, I felt like one of the good guys. What could I do? Other people littered, not me.

Nonetheless, in April 2017, I wrote on my blog about my habits:

> I thought of a new one, especially for city dwellers:
>
> Pick up at least one piece of trash per day from the ground and put it in a trash can.
>
> This one has been formulating in my mind for a while.
>
> New York City is probably cleaner than ever, but its streets and sidewalks are covered with garbage. Most people would probably blame the city for not cleaning it up or not having enough trash cans.

I first visited Manhattan in 1981 and have lived here since 1988. I've seen the city near its grittiest. Did I think my picking up one piece of litter per day would make a difference? No, the idea felt like I'd waste my time, get dirty, maybe sick, embarrass myself, and change nothing. For all I knew, people would see a cleaner ground and litter *more*.

The idea felt like it might fit the pattern that I learned from avoiding packaged food and flying. I saw the problem as a matter of personal

---

15   Philip J. Landrigan et al., "The Minderoo-Monaco Commission on Plastics and Human Health," *Annals of Global Health* 89, no.1 (2023): 23.

responsibility. I found taking personal accountability improved my life. I'd come to live by the phrase, "Don't look for blame but take responsibility to improve things to the extent you can."

I committed to picking up at least a piece a day, still wondering the point, but I started. It took only a second or two per day since, sadly, it's everywhere. The activity opened my eyes to how much more litter existed than I thought. It didn't take long for rewarding feelings to come.

Many pieces of litter I saw prompted me to imagine a story behind them: for example, a mostly-full plastic bottle of Fiji brand water. Someone paid two dollars for water shipped ten thousand miles to a city with water rated among the best in the nation, then only drank a sip. Growing up, we viewed bottled water as a froufrou European affectation. Now we saw it as necessary. The marketers won. The curb showed that people dropped a lot of litter from their cars, often including water bottles with yellow liquid, presumably urine. (I don't pick them up.) I cringed at our depravity.

I came to see just the step of manufacturing something designed to be disposed of as littering, long before a consumer littered it. Once produced, it would reach the ocean. Paying for it funds producing more, as well as for lobbyists and advertisers to promote producing yet more.

The activity forced me to think about my culture. Why would someone litter mere steps from a trash can? Why do they throw away so much unfinished food? Why accept so much packaging and utensils for just a cup of coffee? What happened to sitting to eat or to drink a cup of coffee? Answers to these questions painted a dark picture of our culture. I should be candid: experiencing litter hands-on every day, and the darkness it reveals about our culture makes me want to give up every day—that people could litter so much, that corporate leaders could choose to create so much packaging they know will become poisonous, deadly waste, that billions of people could so complacently accept and pay for more of it. These considerations make me feel futility, despair, disgust, hopelessness, and the like. As much as I don't like feeling these emotions, they are powerful. Every day, I channel their power toward resolve, enthusiasm, and other emotions. For years I had to do so solo. Now I can rely on the friendship and teamwork of the growing alumni community from the workshop.

Most people see litter as a sanitation issue. I don't anymore. It's a too-much-production issue. No sanitation system can keep up with our production. New York City's annual sanitation budget is approaching two *billion* dollars and we're losing. Most plastic comes from fossil fuels, and we produce so much fossil fuel that the cost of plastic has become nearly negligible. Americans view sanitation as an entitlement (somehow that large fraction of Americans who oppose socialized medicine stay silent on socialized sanitation). People once predicted we'd have energy too cheap to meter. Instead we have plastic too cheap to charge for. Restaurants give people disposable dishes even to eat in-house. We don't account for the mess shipped overseas. Why do Americans fight to keep health care from becoming socialized but accept socialized sanitation? Litterers think, "I pay my taxes. Someone should clean this mess, not me," but they're causing taxes to grow.

I went to a supermarket to survey informally the contents of one hundred shoppers' carts. The results? Every item in every cart was packaged. I did a quick calculation. If all eight billion people on earth contributed one piece of plastic per meal that took one thousand years to break down—whether littered or making it to a landfill—we'd create eight quadrillion pieces of plastic. Manufacturing each piece polluted and depleted. Each will become microplastic. They will kill wildlife. It will enter your bloodstream. We may be creating millennia of birth defects.

My action created ripple effects. I held business meetings outdoors during the pandemic. I invited executives and elected officials I met with to join me in picking up litter. After one such meeting, a CEO told me he saw his daughter picking up litter. He asked, "How did you start picking up litter?"

She replied, "Because you do, Daddy."

He told me he started picking up litter after meeting with me. It warmed my heart (though I'd prefer a world without litter for children to pick up).

Another effect came from an executive at a major oil company I coached at the time. We worked on a presentation to introduce our sustainability leadership proposal. He included a picture of his young daughter in a park, smiling proudly, showing off something in her hand. He explained that they had started picking up litter together. He took this picture after she jumped off a jungle gym and chased the piece of litter.

He included the picture in the corporate presentation to show sustainability leadership meant more than corporate results. It meant executives in the firm could improve their own lives and their families' by acting in the field of sustainability.

Starting the Podcast (2017)

I first appeared as a guest on other people's podcasts in 2015 as I prepared to launch my book *Leadership Step by Step*, then continued to appear more often after launching the book in 2017. Some hosts suggested starting my own podcast, but I couldn't think of a topic.

When I had coffee with my friend Jay, I told him about the unexpected rewards of picking up litter. He said, unprompted, "I'm going to pick up ten pieces of litter a day for thirty days." Thirty days later, I asked him how it went.

He said when he started he felt embarrassed, but by the end, he *wanted* people to see him. He enjoyed it enough to commit to more. After some research, he committed to eating less meat—a nontrivial commitment since he lifted weights and had to keep his balance of macronutrients.

His enthusiastic commitments surprised me. All I'd seen of people trying to influence others to pollute less were based on trying to get them to deprive themselves of something they liked. Few sacrificed. Rarely did anyone express joy, let alone do more and want to do yet more.

Jay's result pointed me to a podcast strategy. I would motivate guests to try a small action like ours, expecting them to experience the unexpected reward we did.

I launched the podcast in November 2017. I've averaged 2.5 episodes per week since. I started with several well-known guests, including bestselling authors like Dan Pink 🎤 and Marshall Goldsmith 🎤, Pulitzer Prize-winner Elizabeth Kolbert 🎤, and Super Bowl-winner Bryan Braman 🎤. After a few dozen episodes, I found a flaw in my strategy. I didn't realize it and didn't intend it, but I was trying to convince, cajole, coerce, and seek compliance—what I now call *4C-bludgeoning*. I didn't see the flaw at first because some guests complied with my suggestions and experienced the results I hoped for. Several resisted though, as I unwittingly led them into the results 4C-bludgeoning creates: reinforcing the beliefs that led them not to act in the first place.

Based on my experience teaching and training leadership, I began asking them about their experience in nature to evoke their intrinsic environmental values, then connecting their already-existing emotions to a task of their choosing. They'd often put their guards up, maybe wondering why I asked. Each experience each person shared was unique. All were meaningful, often about immersive experiences in natural environments—mountains, forests, beaches, parks, with animals, and so on.

As I learned to listen actively and supportively, they shared more. I learned to ask them to share what they felt from all five senses. They tended to share emotions of awe, wonder, freedom, calm, peace, connection, and other rewarding emotions. They often sounded inspired to act on the tasks they came up with. When I asked if they were doing it for me, they usually said no. They were doing it for themselves. When I asked how the activity went afterward, they almost always found it more rewarding than expected. Like Jay, most did more than they committed to.

As I shifted from 4C-bludgeoning to inspiring, I noticed my technique followed what I taught in *Leadership Step by Step*. Until then, I had seen sustainability as so big that it would require more than basic leadership. But the basics work, just applying them to sustainability requires fine tuning.

Developing the Spodek Method (2018)

All those podcast conversations led my technique to evolve into the Spodek Method. It is a leadership process to evoke people's intrinsic motivations relevant to the environment, invite them first to think of a way to act on them, then act and share how it went. Because it inspires with intrinsic motivation instead of imposing obligation with extrinsic motivation, it's the opposite of 4C-bludgeoning. It creates a mindset shift followed by continued improvement. It's like learning to play an instrument: playing one scale doesn't do much, but people who play Carnegie Hall play the most scales.

My greatest discovery from the Spodek Method is that everyone has deep, powerful emotions relevant to nature that can motivate them to live more sustainably and enjoy the process.

Before and after the mindset shift is a world of difference—a cusp like the Continental Divide, where on one side a drop of water will reach the Pacific, on the other the Atlantic, a continent away. Before the mindset shift, action feels like deprivation, sacrifice, burden, and chore, provoking hopelessness and futility. After it, acting feels liberating and leads to enthusiasm and compassion.

My Continual Improvement (2018 to Present)

Many people can't believe I dropped my impact by 90 percent in Manhattan or that famous people return for multiple episodes on a sustainability leadership podcast. In some ways, no one is more surprised than I am. My guests' actions inspired me. I felt community growing. Next thing I knew, *The Daily Show* aired a segment on me.

My first steps of avoiding packaged food and avoiding flying were the hard ones. They shifted my mindset. Everything since has been continual improvement, even commitments that may look bigger. For example, I was prompted to unplug my fridge by an article describing how most of the world doesn't refrigerate as much, but I really did it because of my earlier results. By this point, I expected that unplugging my fridge would create unexpected reward too. I knew I wasn't going to die so I walked over and unplugged it. At first I panicked, which made me feel foolish since people lived without fridges for 250,000 years. Many do today. Amazingly, I made it three months my first time trying. I tried again and made it over six months the second time. I aimed for eight months the next time, but made a year, decided to keep going, and may never plug it in again.

Among the results: contrary to my expectations, my food is fresher, I don't need to shop as often (frozen pizza needs refrigeration, not apples), I waste almost no food, and by shifting spending to co-ops and farmers markets, I help them grow, improving access for people without access to fresh produce.

More continual improvement: unplugging the fridge led me to wonder what would happen if I unplugged the whole apartment. Would it be like holding my breath, where I have to breathe extra after? I went to the circuit breaker and disconnected my apartment for twenty-four hours. When I plugged back in, nothing had broken. I didn't need extra energy to make up for lost time.

More continual improvement: I tried disconnecting my apartment for a month. I had no idea how to make it past a day or two. It seemed an opportunity to try solar, which raised challenges. My building wouldn't allow me to install a panel. I posted to my blog for help with solar, but no one responded. But I learned from Teddy Roosevelt: "Do what you can with what you've got where you are." On Craigslist I found used equipment designed for camping and bought it to experiment. Would it provide enough power? Did my apartment receive enough sunlight? How long would charging take?

An early experiment showed the battery could power the pressure cooker to cook a load of stew. That experiment became my first attempt. I disconnected the main circuit and wondered how long I'd make it.

Nobody is more surprised than I am that I made it past a few days, let alone the month, let alone to reach the winter solstice. Once I reached the shortest day of the year, I knew each next day would provide more sunlight and I could make the rest of the year. Eventually I canceled my account with the electric company. Who knows how long I'll make it? Maybe I'll never plug in again. Maybe something will break and I'll have to reconnect to the grid. Either way, over four billion people live in cities. Many lack electric power, but the billions who do impact far more. If those who impact the most can't imagine living without a grid, they won't try. As word spreads, I predict people more resourceful than me will find ways to outdo me. Every human for 250,000 years until a century ago lived without an electric grid. People call me extreme, but I see myself as living more traditionally.

An additional benefit to using less power is living with more *resilience*. Refrigerators are one of our main reasons for "needing" our power grids to stay up over 99.9 percent of the time. Power companies building early grids promoted refrigerators to require power nonstop so they could keep their plants running nonstop. That need requires overbuilding and redundancies, which increase costs, taxes, and pollution, and decreases national security. Nearly a third of Americans have two refrigerators. I have a friend with four in just one of his homes. Resilience to where everyone could handle a day or two without power could reduce costs, taxes, and pollution, increasing national security even if no one used less power.

Among the results of disconnecting from the grid: I save money, am distracted less by social media, procrastinate less, and have more free time. I do things our ancestors did that require less power, like writing, singing, and volunteering.

Developing a Strategy: Top-Down, Bottom-Up, Everywhere All at Once, Starting Here and Now With You and Me (2021)

I realized any effective sustainability leadership strategy would have to be top-down and bottom-up, so I started calling my strategy "top-down, bottom-up, everywhere all at once, starting here and now with you and me." When the process is joyful and leads to a brighter, desirable future, the more we learn it, the more we share it.

Working with world-renowned influential leaders is the top-down part of my strategy to help lead their organizations. Taylor Swift, the CEO of Exxon, and presidents of nations influence tens to hundreds of millions of people. The Spodek Method leads them to share their intrinsic motivations, commitments, difficulties, ignorance, and other vulnerabilities, which gives them integrity, credibility, character, and other leadership qualities.

The bottom-up part of the strategy is teaching the Spodek Method to participants as well as coach-training so it can spread without me as a bottleneck. It works when people act on intrinsic rewarding motivation, not from 4C-bludgeoning. I envision the workshops growing like Weight Watchers, Alcoholics Anonymous, and CrossFit. In those cases, people go through tremendous personal growth and can start new groups to bring their transformation to others.

Workshops (2023)

I'd been teaching leadership and entrepreneurship classes at NYU for nearly a decade and practicing the Spodek Method on the podcast half that time, but organizing and leading independent workshops were another endeavor. Various people had volunteered with me on several small projects, but this one required significantly more coordination.

Friends stepped in to help. We organized and created the first workshop. I feared people might not show up after the first session. Instead, everyone finished. One graduate, Evelyn 🎙, became a teaching assistant for the next two cohorts, then taught her own sessions. You read her mother Beth's recommendation on page 20. More and more alumni are training to become coaches too.

What My Experiences Taught Me Overall

It's natural to see our environmental problems and solutions as scientific, technological, market-based, or political. We learned about them from scientists. Innovators, businesspeople, and politicians say they can solve them, but the problems and solutions are social. The environment is responding to our behavior, which results from our culture. To change the environmental results, we have to change our culture.

Looking back, it's obvious: If I wanted to adopt another culture, say French, it works much better to go there and learn from experience than to try to learn it from books from people who never visited France and only visit when I think I have reached fluency. I studied French for years in school but learned more in two weeks living there—particularly from making mistakes and from interacting with French people. Likewise with sustainability: **We as individuals and nations will learn more by living sustainably ourselves than from books by people living unsustainably. We will learn more from making mistakes and from interacting with sustainable people than trying to solve too many problems before starting.** Since nearly no one lives sustainably today, we can restore sustainability fastest and most enjoyably by forming groups of people transitioning together.

CHAPTER 4

THE SPODEK METHOD

Before the mindset shift, lecturing people with facts and numbers, telling them what to do, scaring them with "Code Red" warnings, and other 4C-bludgeoning tends to reinforce their beliefs driving the undesired behavior. By contrast, many people who have practiced the Spodek Method feel gratitude and inspiration.

The Spodek Method in practice amounts to two conversations, often under thirty minutes each, where you lead someone to share their environmental values, then a way for them to act on them, then after they do it, to ask how it went. Like playing a song on a piano, learning it takes practice, but then it becomes easy. Everyone plays songs slightly differently and every person you do the Spodek Method with will respond differently.

The how-to workbook teaches the basics. The basic online course provides a few videos walking you through how to do it, starting with the easy parts and working up to the challenging parts. Each video gives you an exercise to practice. After you practice, you write your experience and reflections on a message board that lets you see the reflections of everyone who has done the exercise before, often revealing insights and angles you may have missed. The alumni community is growing.

The table below expands on the difference between before and after that mindset shift. The Spodek Method is the only way I know of that creates these results.

BEFORE		AFTER
Guilt, shame, helplessness, hopelessness, dread.	→	Freedom, joy, fun, connection, community, meaning, purpose.
Rationalizations like: "What I do doesn't matter," "Only governments and corporations can make a difference on the scale we need," "The plane was going to fly anyway," "If we don't progress, we risk reverting to the Stone Age or reaching a *Mad Max*-like apocalypse."	→	Discoveries like: "Nature is worth saving," "I wish I had started earlier," "I can't believe I missed this problem and solution staring me in the face all along," and "What have we done?"
Abdication, resignation, capitulation.	→	Personal responsibility, compassion.
"I have to."	→	"I get to."
Focus on extrinsic motivation.	→	Focus on intrinsic motivation.
Abandoning basic human values embodied in *Do Unto Others As You Would Have Them Do Unto You* (Golden Rule), *Leave It Better Than You Found It* (Stewardship), *Live and Let Live* (Common Decency).	→	Restoring the Golden Rule, Stewardship, Common Decency, and other basic human values.
Destroying others' lives, liberty, and property with impunity.	→	Respecting others' lives, liberty, and property.
Don't want to learn more about nature or environmental problems; it makes you feel guilty.	→	Want to learn more about nature and our situation; it motivates action.

"Here are ten little things you can do to help the environment."	→	"Here's how to connect with your emotions and motivations on the environment and stewardship."
"If enough people do little things, they can all add up."	→	People you lead feel gratitude so they want to lead others to lead others and so on, spreading on its own and adding up.
"Better many people do a little imperfectly than a few people be perfect."	→	Better to focus on intrinsic motivation and leadership skills so people continually improve and help others to improve too.
"I've done my share. I'm one of the good guys." "I'm doing all I can." or "Aren't I doing enough already?"	→	"What more can I do and whom else can I engage?"
"Climate change is complex enough to understand or solve. All our other environmental problems are too numerous and complex to understand or act on: deforestation, ocean acidification, fisheries collapsing, extinctions, biodiversity loss, and so on. I throw up my hands."	→	"Our environmental problems all result from our behavior, which is driven by our stories, beliefs, images, role models, and what constitutes our culture. The difference between the fertile environment of our ancestors and our environment today is the physical manifestation of our culture and values."
Sustainability means deprivation, sacrifice, burden, and chore.	→	Sustainability means freedom, fun, joy, community, connection, meaning, purpose, connection to nature, and more.
Nature is scary and threatening.	→	Nature is abundant and nurturing.
Human ingenuity requires more energy.	→	Humans are ingenious with or without energy.

"Others should change to pollute less, but what I do is so important, I should still pollute."	→	"You can't lead others to live by values you live the opposite of."
Discouraged.	→	Enthusiastic.
"Tomorrow will be worse than today."	→	"I can improve tomorrow."
Outraged at others causing the problem.	→	Compassionate for others.
Either we save ourselves or we collapse.	→	There are levels of disaster. Everything we do can lessen others' suffering.
"Thinking of how my behavior affects others is a burden."	→	"Thinking of how my behavior affects others is one of the best parts of being human."
"I have to balance caring about nature with living a good life."	→	"I want to balance my comfort and convenience with how my behavior affects others."
"Technology, markets, and laws are our only hopes, or at least our best hope."	→	"Technology, markets, and laws augment the values of the people using them. If we don't restore values of stewardship, they will accelerate our trajectory."
"I am powerless."	→	"I am powerful."
Don't want to talk about it.	→	Want to share experience.
Want more energy and power.	→	Enjoy experiencing nature and time with people.
"There's no point. Nothing will make a difference."	→	"I wish I had started earlier."
Craving.	→	Calm.

Twisted up inside from mental gymnastics.	→	Seeing how it all fits together.
Insecure.	→	Secure.
"People who act are stupid and wasting their time."	→	"People who aren't acting are suppressing and denying."
Addicted, dependent.	→	Free, independent.
Doof (see chapter 5).	→	Fresh, local produce.
"The people causing the problem are (unrepentant capitalists sleepwalking into destroying the environment and enslaving people) or (ignorant socialists or communists sleepwalking into global totalitarianism we won't be able to undo)."	→	"We're all contributing to the problem and we can all help solve it."
"We must destroy the system."	→	"We must connect with our deepest, most human values. That's what drives the system so that will change the system."
Resignation, giving up.	→	Passion, resolution.
"Others are hurting me. I'm a victim."	→	"I can help others."
"I am entitled to the life I planned before, no matter what."	→	"I acknowledge the life I planned before depended on hurting innocent people, whether I liked it or not."
Spread facts, numbers, instructions.	→	Lead others to experience mindset shift.
Fear, anger.	→	Love.

Want to know less of the problem so you can avoid facing it.	→	Want to know more so you can solve it.
Charts, graphs, numbers.	→	Problem-solving.
Analyzing and planning.	→	Doing.
Alone.	→	Together.
Can only envision leading to dystopia or Stone Age.	→	See brighter future of freedom and equality.
Blame.	→	Responsibility.
Avoiding emotions from internal conflict.	→	Examining and resolving conflict.
Follow.	→	Lead.
"Balancing" sustainability against other values.	→	Recognizing they don't conflict.
Comfort, convenience, and procrastination.	→	Meaningful engagement.
Hoping someone will act later.	→	Acting now with confidence.

Results

Here are four results from podcast guests I led through the Spodek Method.

Colonel Mark Read

I call Colonel Mark Read 🎤 my one-man wrecking crew after his experience on *This Sustainable Life*. Mark heads a department at the US Military Academy at West Point—one of the world's premier leadership training institutions, making him a leader of leaders.

On the podcast, he committed to reducing his household garbage by half for one month. I asked how it went after that month. His family supported the challenge, but the month was December 2018, meaning Christmas, gifts, and more trash if they didn't do something different. They met to solve how to handle the challenge. They first decided to skip wrapping paper. They continued and decided to skip material gifts, eventually deciding together to take a staycation, without distraction, at a nearby hotel. Instead of material gifts, they gave each other time and attention.

Did the kids rebel or feel they were missing out? He described it as their best Christmas ever—not *despite* foregoing gifts but *because* no gifts distracted them. The change stuck. In holiday seasons since, he's told me his children look forward to their material-gift-free Christmases together, preferring it that way.

I call Mark a wrecking crew for wrecking the most common excuses pre-mindset shift people give me for not acting. He turned each excuse into an advantage, making his family closer, his work more effective, and America safer.

The top four excuses I hear are: family, job or income, culture, and tradition or religion. Mark loves his wife and kids as much as anyone. He wouldn't sacrifice time with them. He loves his job defending his nation, to the point of having committed to risking his life if necessary, and teaching cadets the same passion. Regarding culture, many politicians say that America is about more and bigger, especially material stuff and the GDP. Well, nobody loves their country more than Mark loves his. Since his commitment, he shares with cadets how his experience reducing waste contributes to their mission, not detracts. Regarding tradition, modern America has reduced much of Christmas giving to material gifts, but Mark's family is restoring spiritual giving and humility.

Andrew C.

Andrew 🎤 didn't follow politics that much growing up, but during the 2016 election, Donald Trump electrified him. He supported Trump in 2016 and more in 2020. He became the co-host of a show on MAGAMedia.org that evolved into *After Dark with Rob and Andrew* on the America Out Loud Podcast Network. His bio there describes him as "a social media pundit, writer, and podcast host. He believes in traditional conservative values, including the right to free speech, America First, limited government, and the Bible."

Andrew's connection to the environment stuck with me. Most people connect with raw nature. He connected to small-town America. He loves his town in Illinois. He doesn't like how big cities are dirty and growing more so. He sees litter overrunning his town.

When I invited him to act on his values, he thought of his Gatorade habit. He drank four bottles a day, creating a lot of trash. He came up with the idea of recycling them for a month and collecting a return deposit. As someone who hasn't bought a plastic bottle in a decade, did I suggest he do more? Did I debate the effectiveness of plastic recycling? No, because I saw him acting on his values. Big or small doesn't matter so much as meaningful. I knew from experience what would likely happen.

A month later, when I asked him how it went, he told me he found the process fun. His girlfriend joined him in it. He noticed his trash load decreasing. He recounted how in getting money back for recycling, the small amount wasn't the point. It just felt good. He remarked how easy it was and that everyone should do it. Why? Not because I told him to do something or told him new numbers or facts, but because it delivered on his intrinsic motivation: it decreased the garbage in his small town.

Rhonda Lamb

I met Rhonda 🎤 at a potluck lunch when we sat next to each other by chance. She was there with her son. We chatted. I learned she was a single mom from the Bronx. Since people have responded to my stories of personal action with variations on, "You don't know what it's like to be a single mom in a food desert," here was one I could learn from.

I invited her to the podcast. As I do with local guests, we recorded at my place where I made my famous no-packaging vegan stew. She loved it for the taste, simplicity, and accessibility, which led us to arrange for me to lead a workshop to show her community what I did.

That Bronx workshop was my last public event before the pandemic. I've been to farmers markets in the Bronx and seen residents pass beautiful, cheap farm-fresh produce. I wanted to help increase demand. I brought all the ingredients and equipment—knife, chopping board, pressure cooker—to make a stew at the community center church Rhonda had found to host the event.

After I finished, the first person said, "It's nice what you do, but we can't do it here. We don't have access like you do to things like nutritional yeast."

The next person said she knew where to get some.

The third person said she knew where to get other supplies and the group started talking. As they solved problems together, Rhonda said, "Josh, you planted the seed. It may take time, but now we know we can do it."

As it happened, one of my first public appearances after the pandemic was another Bronx workshop Rhonda organized, this time at a public park. Drew Gardens is a former industrial site that community volunteers cleaned and turned into what I consider one of New York City's most beautiful spots. It illustrates something I can't prove, but seems to happen every time: acting on environmental values leads to community and connections. The bigger the challenge, the more solving it involves other people and leads to collaboration and connection.

Lorna Davis

Lorna 🎤 served as CEO of Danone Wave (now Danone North America), where she led the transformation of the six-billion-dollar organization into the world's largest B Corp, though still a global "leader" in producing packaging (it produces no products I would buy). She retired before appearing on *This Sustainable Life* so she appeared as a private citizen. On the podcast, she committed to buying no new clothes for a year, one of the longest commitments by a guest.

I didn't tell her, but I was skeptical she'd make it. A couple months in, I asked her how she was doing. My skepticism was misplaced. Beyond not buying clothes, she examined her wardrobe and got rid of clothes she didn't wear, freeing space in her closet and mind. The resulting freedom prompted her to examine her calendar and cut back on appointments and relationships also taking up space.

Over the course of the year, she updated me on more changes. Once, a boot strap broke on the way to a board meeting. Before, she would have bought new boots. This time she strode into the meeting with the boot loose. She told her colleagues how it broke and her commitment. A team exercise in problem-solving emerged, creating joy, fun, and connection. Another time she texted for advice on what to do with unrequested soy sauce packets that came with her sushi. The point isn't the packets, but her fun and freedom, which became infectious. Other executives followed her commitment with commitments of their own.

Midyear, she was invited to give a TED talk on its main stage in Mumbai co-hosted by the prestigious Boston Consulting Group. She told me that the old her would have first wondered and stressed on what to wear. Instead, she took friends' advice to wear what she liked most. She did and delivered comfortably. Her talk was approaching three million views at the time of this writing. She also continued her commitment a second year.

Stephen M. R. Covey

Stephen 🎙 is the bestselling author of *The SPEED of Trust* and *Trust & Inspire: How Truly Great Leaders Unleash Greatness in Others* (and son of the author of the blockbuster *The Seven Habits of Highly Effective People*). After experiencing the Spodek Method, he said the experience recalled a quote attributed to Antoine de Saint-Exupéry, author of *The Little Prince*: "If you want to build a ship, don't drum up the men to gather wood, divide the work, and give orders. Instead, teach them to yearn for the vast and endless sea."

He added that the Spodek Method

was [an] emotional experience. It was great. It felt aligned and appropriate to what I was doing . . . That was inspiring to me. I love this whole approach and I've been thinking about the [Saint-Exupéry] quote. That's an inspiring pull approach rather than a push strategy. That's the kind of leadership that's really effective.

You're inspiring people, not just motivating them.

Oliver Burkeman

Oliver 🎤 is the bestselling author of *Four Thousand Weeks: Time Management for Mortals.* He said of his experience with the Spodek Method:

The idea that it's actually an act of kindness toward yourself to do that is fascinating to me. That's what I get from what you're saying: it's not actually a question of having to choose between self-gratification and being good to the planet.

It's not that people have to exactly copy you. It's that people would discover something very rich, full, and meaningful in life by approaching limitation in the same way . . . It's about living more consciously in this world of limitation and trade-off and seeing what that opens up and how that liberates you.

You're right to talk about how quickly these rewards come from voluntarily stepping into this.

Tony Hansen

Tony 🎤 is a director at the management consulting firm McKinsey. He said of his Spodek Method experience:

You opened some doors. The idea was there but I'd come up with excuses for why I couldn't engage now. If [I'm] honest I'll be a whole lot more effective right now . . . than I might be in fifteen years time. It makes a huge amount of sense to do right now so I thank you . . . because I don't know if I would have acted on it. Now that I've committed to it, I will.

Very few have done what you've done.

He continued:

[Those] not doing it: one, don't recognize what it takes, two, don't recognize the benefits of it, and three, can't credibly convince others.

There's no better way than trying it yourself. You can then speak with authority and awareness, as opposed to just saying oh we should do this but not really intending to.

Sometimes [we] require some form of awakening that . . . gives intrinsic motivation to do something, something different . . . through that action of doing something differently, you can build momentum.

The Spodek Method is one of those tools to enable that awakening.

Practicing the Basics

One of my favorite videos online may be one of the most boring. It shows LeBron James practicing for an hour with a trainer. If you've seen him play, you've seen his nearly superhuman feats. He doesn't practice what makes the highlight reels, though. He practices the basics. Over and over, he practices the basics. When he practices them enough and doesn't have to think about them, he looks like magic.

The point of my experiments isn't my individual impact. I'm practicing the basics, learning what's possible. I'm developing the skills and experience to help others. I'm doing what all performers do to achieve mastery. LeBron's field is basketball. Mine is sustainability leadership.

Admiral and Navy SEAL William McRaven gave a speech to graduating seniors, "If you want to change the world, make your bed."[16]

"Wait," you might object. "Isn't it his job to defend the nation? What difference would making my bed make?" He knows that making your bed helps create mastery and excellence. It's practicing basics. As it happens, I make my bed every day. I meditate and exercise every day too. These activities contribute as much to sustainability leadership as unplugging my apartment, maybe more. They enable me to live by my values. Only with them can anyone speak with integrity, credibility, character, relevant experience, vision, and other properties of effective leadership.

Virtuosity and Freedom

Before mastering fields that require skills like sustainability and sustainability leadership, things look black and white: either humanity saves itself or collapses. With mastery we learn nuance, subtlety, and complexity. Everything we do that affects the environment affects everyone. When you know what works and how to do it, action becomes the antidote to helplessness, hopelessness, guilt, and shame.

Sustainability is consistent with your values whatever your background and leanings—conservative, liberal, libertarian, old, young, rich, poor, or however you identify yourself. Whatever your skills, the movement needs you. Whatever you give, you will get back more. It will increase your freedom, joy, fun, self-awareness, and love.

I've led thousands of people through it. Learning the Spodek Method through practice after being 4C-bludgeoned is like hearing music after only being lectured music theory. Continuing to practice is like singing or playing an instrument yourself. You express yourself in new ways. It's

---

16 "Navy Seal William McRaven: If You Want To Change The World, Make Your Bed!" Goalcast, August 17, 2017.

liberating. As the great dancer Martha Graham described:

> Your goal is freedom. But freedom may only be achieved through discipline. In the studio you learn to conform, to submit yourself to the demands of your craft, so that you may finally be free.

Or as Navy SEAL Jocko Willink put it, "Discipline equals freedom."

"But My Life Is Fine. Why Should I Engage?"

As more people asked me about disconnecting from the grid, I had to search to explain why I engaged so much. I had to climb eleven flights to put the solar equipment on the roof, often twice a day. If people didn't understand why, they'd see me as a quirky guy or zealot. I may be quirky, but I sensed I was motivated by universal human values. Climbing stairs and using solar was a distraction. But I couldn't at first identify my deepest motivations to communicate them.

At first, I thought I might be feeling what Dr. King meant when he said "unearned suffering is redemptive." But I wasn't suffering. After a few more weeks, I wondered if I was doing it for the exercise since I would count climbing stairs as exercise on cardio days, but I could tell I wasn't doing it for exercise. After a few months I wondered if I was doing it as a ritual. I have many rituals. I value that they raise my self-awareness. But again, on reflection, I saw I wasn't doing it as a ritual.

The answer came after nine months from an unexpected direction—the memorial service of a mentor of mine, Frances Hesselbein 🎙, at a church in midtown on Park Avenue. Frances had died in December at her home in Pennsylvania at 107 years of age. I had met her nearly a quarter century earlier when she spoke at a business school class. She became a mentor.

Frances rose to prominence as CEO of the Girl Scouts from 1976 to 1990. She was the main person transforming them to reflect the nation's diversity, helping girls become anything they wanted, not just housewives. She was widely recognized for her leadership ability including President Clinton awarding her the Presidential Medal of Freedom, renowned business author Peter Drucker calling her "the best leader in America," and West Point appointing her to an endowed chair in its leadership department.

She's known for her catchphrase "to serve is to live." Several eulogists quoted another favorite of hers, "work is love made visible," which quoted the poet Khalil Gibran. I happened to sit behind the head of West Point's leadership department, Colonel Everett Spain 🎤, and his colleague, Colonel Katie Matthew, both in their dress army ranger uniforms. The quote prompted me to ask them about love, which I wouldn't usually think of discussing with soldiers. Nonetheless, they spoke of the love soldiers feel for their country, those at home they protect, and fellow soldiers they trust their lives with. I appreciated what they shared but didn't think more of it in the moment.

After the memorial, perhaps thinking of Frances, I remembered a time that may seem unrelated at first. An elderly woman stopped me near my building to ask if I could help her cross the street. I was in a hurry, but put my arm out. She gripped it and I helped her across. It was no big deal. It took maybe two minutes. I felt elated from helping her. I felt even better for having been in a hurry.

By chance, not much later, I read a story of Jocko Willink in his book *Extreme Ownership*. In Afghanistan, on patrol with less-trained Afghani soldiers, they were fired upon. The SEALs turned to face the attack. The Afghanis fled. One was wounded and fell. His comrades left him in the line of fire. When the SEALs saw him, they ran into the line of fire to bring him to safety.

I normally don't think of men, particularly warriors, as talking about feeling love for each other. I don't pretend to know what combat feels like, and an elderly woman crossing the street sounds incomparably different, but I saw a common thread through these stories. They involved helping the vulnerable and innocent, even when we don't have to, even when we struggle. It's deeply human to feel good *because* we struggle for what others need.

It took a few more weeks to process the meaning of "work is love made visible" despite having heard Frances say it for decades. I saw climbing all those stairs in a new light: I was not doing it for myself, my health, or some abstract concept of "the environment." I was doing it out of love for my fellow humans, in particular innocent ones that my polluting and depleting would cause to suffer. I also think of the billions of internally conflicted people polluting and depleting who ache to stop but lack leadership, role models, and a vision of a brighter future.

Alleviating suffering through personal action is not deprivation. It's love. Frances told me for decades: *work is love made visible*. She was quoting Gibran, but the sentiment came also from luminaries from countless cultures.

Had I not lived this experience, I might think this talk sounds like someone I'm not—hippie, tree-hugger, new age, or the like. I don't know what else to say, though, because what motivated me was love borne of deliberate, purposeful effort to live by my values when it was easier to procrastinate.

I always knew something was wrong when "extreme" and "admirable" became the most common adjectives people called me. Both terms set me apart from them. People paint me as other or different—at least they want to. I let the descriptions stand for a long time, but something wrong about them lingered. I couldn't put my finger on it.

Frances's memorial led me to another realization: **I wasn't acting on something extreme, quirky, or unique to me. I was acting on the most universal of human emotions and motivations. Moreover, I wasn't helping "the environment" in the abstract; I am helping people. I'm loving my neighbors as myself. I'm not just trying to avert future catastrophe. I am alleviating suffering and subjugation here and now.**

I'm doing the most universally human behavior given the world we were born into. This realization revealed why even when people didn't understand me when I felt too tired to climb the stairs, when my torn meniscus hurt, or whatever problem emerged, nothing in me considered giving up. Giving up would mean not comfort but irresponsibility, cruelty, and lack of integrity.

"I'm Pointing at the Moon. They Keep Looking at My Hand."

The New Yorker published a letter to the editor I wrote in October 2022 commenting on an article about Americans and Europeans trying to help reduce food waste in Africa by bringing them refrigeration. I wrote that we waste more food than they do, we pollute more doing it, and our diets are less healthy. I wrote a way we "could help people in developing countries is by reforming our own system so that we pollute our shared world less. We could make it easier to run farmers markets here, thereby reducing our overreliance on refrigeration."

The New Yorker is prestigious. Why did they publish my letter?

Probably what I wrote next: "Cutting back on fridge use might seem difficult, but it is possible. I speak from experience. After reading a 2017 article about how other cultures refrigerate less, I tried unplugging my fridge. I didn't think I could make it a day, but, using a combination of tactics, including fermentation and effective shopping, I made it three months. My next try lasted nearly seven months. Now, on my third try, I'm a week short of making it a full year."

I wrote from the center of the Venn diagram of sustainability leadership—that is, with integrity. Being published there led to another outcome that infuriated me, at least temporarily.

Experience
Leading

Experience
In
Science

Experience
Living
Sustainably

A reporter from the magazine's "Talk of the Town" section asked to profile me. I said I knew that section tended to print look-at-this-quirky-guy pieces. My work on sustainability leadership is mission-driven and grounded in research and experience. I asked that he not cover me as quirky but instead cover the mission. He agreed, or at least that's what I remembered.

He visited and we spoke for hours. I shared my mission, strategies, Spodek Method, and results. He interviewed three corporate leaders I worked with, from polluting, influential companies: Shell, Exxon, and Boston Consulting Group. Each described my work as essential, effective, engaging, and that nobody else was doing it. I felt he had a great opportunity and platform to help on an issue the magazine covers and that affects him personally.

Instead he wrote a look-at-this-quirky-guy piece. I lamented the lost opportunity. The world could have learned of an effective, business-tested path toward sustainability leadership. Eventually my response turned to compassion. It wasn't his responsibility for me to be understood.

Calming down allowed me to reflect on journalism covering sustainability in general. Even among writers who understood our environmental problems, nearly all of them told stories but didn't motivate action. I came to see journalists like people passing a child drowning in a pond who, instead of saving the child, put their profession first and write a story about it. The story is compelling so people read it. Other writers notice a conveyor belt is putting more babies in the pond and they're all drowning. They write about the kids, the conveyor belt, and so on, but continue not saving the children.

In time I found another perspective on how to describe how they covered me: I'm pointing to a brighter future, but they keep looking at my hand. *The New Yorker* wrote how I have dirt under my fingernails.

PART 2

OUR WORLD

CHAPTER 5

DOOF AND ADDICTION

My path toward stewardship began with food—in particular, avoiding packaging—and food remains a major focus, but I found a problem deeper than litter, greenhouse gas emissions, or factory farming: our culture of addiction. To clarify, I'll explain the concept of *doof* since the clarity it brings will help us understand our culture and its effect on the environment.

Doof

Avoiding packaged food led me to read a lot of books on food and nutrition. Michael Pollan's book *Food Rules* says, "Eat food, not too much, mostly plants." By "food" he means not industrial products designed to resemble food, which he lacked a word for. Food writers use many terms to distinguish these non-food products from food, like "ultraprocessed food," "junk food," "fast food," and "comfort food."

Using the word "food" in phrases like "junk food" confuses things they're trying to distinguish. It leads people to say, "Maybe junk food isn't the best food, but it's still food," often continuing, "and a single mom in a food desert with three kids and three jobs can fill their stomachs with more spending a dollar at McDonald's than on vegetables from a store," even though research shows otherwise.[17]

A recent article in the *British Medical Journal* reported that 14 percent

---

17   Brian Sodoma, "Is Fast Food Really Cheaper Than Healthy Eating?" *Forbes*, January 16, 2020.

of adults and 12 percent of children may meet a clinical definition of "addiction."[18] In particular, "Refined carbohydrates or fats evoke similar levels of extracellular dopamine in the brain striatum to those seen with addictive substances . . . Behaviours around ultra-processed food may meet the criteria for diagnosis of substance use disorder in some people."

I'll state it more bluntly: whatever we call it, this stuff addicts as much as drugs. Most drugs are refined from plants and fungi, but we don't call morphine "poppy extract" or cocaine "coca leaf extract." If we did, people might consider them herbal extracts and possibly healthy. A Frito corn chip is as different from corn as cocaine is from coca leaf. They are qualitatively different. Hydrogenated oils are not food any more than morphine is poppy.

Doof—food backward—is the word we've been looking for. Doof means what the terms above mean without using the term *food*. I recommend never calling doof "food." Distinguishing them will change your life. It will reveal to you how much our culture has corrupted what we put in our mouths. People who use the term include bestselling food and nutrition authors Dr. Joel Fuhrman 🎙 and Dr. Michael Greger 🎙, as well as New York City's Mayor Eric Adams 🎙 .

I'm not suggesting never to consume doof. I am suggesting to avoid confusing it with food, because it isn't. If you're hungry and consume doof, you may have filled your belly, but you haven't eaten food. Since it delivers calories but rarely nutrition, you'll either sacrifice your (or your children's) health or eat more food later, rendering that doof empty calories, wasted time, and wasted money. Parents can make their children's hunger go away by giving them meth too, but that doesn't make it food.

Most supermarkets put the produce section at the entrance to try to create an atmosphere of freshness and health. Most of the rest of the store is doof because they generally allocate shelf space by profitability.

18   Ashley N. Gearhardt, Nassib B. Bueno, Alexandra G. DiFeliceantonio, Christina A. Roberto, Susana Jiménez-Murcia, and Fernando Fernandez-Aranda, "Social, Clinical, and Policy Implications of Ultra-Processed Food Addiction," *BMJ* (2023): 383.

Many people struggle with their diets, not knowing what to eat or not. I did. A lot of confusion comes from thinking doof is food, that the "Froot" in Froot Loops involves fruit—as doof manufacturers intend.

Do you feel anxious and confused because products touted as healthy one day are found unhealthy the next? They're nearly always doof. Distinguishing them from food can reduce that anxiety. When you identify them as doof, you don't have to fear them. Actual food rarely causes such concern, barring special needs like allergies.

Doof industries do more than squeeze food from stores. They squeeze resources from neighborhoods, especially money, time, and real estate. People often confuse low prices with helping poor people. McDonald's, Nestlé, Starbucks, Danone, and peers are only helping their shareholders. They cause dependency and ignorance of how to shop and cook. They deprive communities of time, money, and resources while implying they are solving the problems they created. Once a community can't access food, people in it lose their skills to cook from scratch and become *more* dependent. Doof companies design their business models to create that dependency. Buying doof funds them and their lobbyists.

A doof store entering a neighborhood is like a pawn shop, payday loan store, or cheap liquor store. Those stores may offer some value, but they tend to extract money from the local economy, impoverishing them. Likewise with most delis, convenience stores, takeout restaurants like Domino's, or sit-down ones like Olive Garden. The appearance of any of them suggests it will start making that community dependent. Supermarkets including Whole Foods and Trader Joe's identify themselves with fresh produce, but the supermarket I mentioned on page 61 where I found every product packaged was a Whole Foods. They sell mostly doof. Doof-sellers use addictive tactics to lure people in, especially children, and do what they can to push out food options like farmers markets. By contrast, food co-ops, farmers markets, and credit unions offer similar services while keeping funds local.

I'm not a professional dictionary writer, so I haven't written a professional definition for "doof" yet. Mainly, "doof" is what writers mean by the terms like "junk food" without implying they're food. Different people can draw their lines between food and doof differently; we all still benefit from distinguishing them. Some signs to identify doof: it's nearly always packaged, it didn't exist before last century, and it includes ingredients most kitchens don't. But the key distinction between doof and food is the intent of the people making it: if their goal is to produce craving or dependence, it's likely doof.

Studies show that doof triggers the brain's reward mechanisms faster than nicotine and many illegal drugs: "Daily snacking on processed foods [i.e., doof], recent studies show, rewires the brain's reward circuits. Cravings for tasty meals light up the brain just like cravings for cocaine do . . . A 2021 study showed, for example, that people with binge eating disorder exclusively overeat ultraprocessed foods [i.e., doof]."[19]

When you see that doof is designed to create dependence, even addiction, you'll see it everywhere, not just products for your mouth. As you use the term more regularly, you may find yourself applying it in places like phone apps, online stores like Amazon.com, social media, fast fashion, and TV shows designed for binge-watching.

Doof is a defining concept of our culture: creating dependency by manipulating our reward mechanisms while claiming it benefits us. Identifying and removing doof from your life improves it, though it may require effort. It may feel like you're giving up something you need, that improves your life, that makes you feel warm inside.

What Doof Is Doing to Our Culture

The Industrial Revolution happened in waves. The first wave employed some of the greatest minds of its time to use mechanical engineering to reduce labor and extend people's abilities to build, travel, and more.

---

19 Robert H. Lustig, "Ultraprocessed Food : Addictive, Toxic, and Ready for Regulation," *Nutrients* 12, no. 11 (November 2020): 3401.

Later waves employed great minds of later generations as civil, chemical, and computer engineers.

Today, doof industries employ today's brightest minds to use psychological engineering to manipulate: the human reward system. They've developed ways to motivate you more effectively than many of us can resist. We're increasingly addicted, feeling increasingly connected to the objects of our addiction, and making them part of our identities. We even love them.

Doof transcends chemicals, as addiction researcher Bruce Alexander 🎤 described,

> Global society is drowning in addiction to drug use and a thousand other habits. This is because people around the world, rich and poor alike, are being torn from the close ties to family, culture, and traditional spirituality that constituted the normal fabric of life in pre-modern times. This kind of global society subjects people to unrelenting pressures towards individualism and competition, dislocating them from social life. People adapt to this dislocation by concocting the best substitutes that they can for a sustaining social, cultural, and spiritual wholeness, and addiction provides this substitute for more and more of us.[20]

When I talk about substance addiction, I'm talking about this cultural effect. Alexander clarified, "Addiction is much more a social problem than an individual disorder."

Avoid doof long enough and you'll see Fritos and other doof products as closer to heroin than to corn. Likewise with suppliers. You'll also find Nestlé, Danone, and Coca-Cola less like farmers and more like the manufacturers of OxyContin or meth.

We don't need doof. Humans lived without it for 250,000 years. It creates waste, pollutes, and depletes. It leads to litter, noise, and much of

---

20 Bruce K. Alexander's website.

our garbage. We used to sit with family and friends to eat meals together. Now we consume doof while walking. Landfills used to contain only biodegradable items, now we fill them with plastic packaging that lasts millennia. Doof vendors insert advertising into our lives everywhere they can. Many diseases attributed to overeating, like heart disease, diabetes, and strokes, result more from doof than food. I'm not saying that only cutting out doof will stop those diseases. I am saying that, to suggest that eating causes these diseases is like saying breathing causes lung cancer, when *breathing through cigarettes* does. Smoke isn't air and doof isn't food. When we see social media, fast fashion, and other industries that promote craving as doof, we see doof contributes to depression, anxiety, and related problems that are increasing.

The largest and most profitable companies—including Amazon.com, Apple, Walmart, much of the Fortune 500, and especially those from Silicon Valley—increasingly find ways control us the way doof companies do, always looking for more techniques to hook and more routes to inject those techniques. If you can't stop scrolling, checking notifications, buying things you don't need, and related behaviors, entire industries are built on creating that dependence. They don't mind if you suffer results like obesity, sleep deprivation, anxiety, poverty, and lack of free time.

Addiction

I don't use the word "addiction" lightly. In 1985, a story appeared in the *Philadelphia Inquirer*: "Nightmare Of Life With A Coke Snorter."[21] It profiled a family where the son, Seth Witonsky, was using cocaine and the parents, Carl and Pecki, were in denial. I remember it because my mom knew the family. Also because it was a big story about cocaine, uncommon in the news then. *Miami Vice*—a TV series that featured cocaine use and dealing more than any show had before—had just started and college basketball star Len Bias hadn't yet died from cocaine use. I see it as a

---

21 Dick Polman, "Nightmare of Life with a Coke Snorter," *Philadelphia Inquirer*, January 6, 1985.

story of our times, only today it applies to a lot more behaviors that are legal, accepted, and promoted—and profitable.

Some excerpts:

> At dawn he would lie immobile, like a log on a stagnant lake, cast adrift by his own inertia. It's a new day, and everything stinks: sunlight, school, Mom hassling me in the kitchen . . . I'll have to scare her, swear at her, get her off my back. Everybody's on my back . . .

> Out in the kitchen, Pecki Witonsky would be girding for the worst. She hated to set foot in her son's room; she called it "the devil's den." Sometimes she'd do it anyway, but Seth would always greet her with a choice remark, such as "I'll bash your head against the wall if you don't get out of here." . . .

> Several months after his bar mitzvah, Seth stole his father's sports car. Carl Witonsky was away on business, and his wife had gone for a walk . . .

> Carl Witonsky returned from his trip but was not concerned. As he recalls, "When Seth stole the car, I thought, 'Why would anyone throw him in the clinker for that?' I almost thought it was okay, a macho thing. I didn't believe he was using drugs. So to take your father's car out for a spin, well, I'd rather have that than have him sitting up in his room doing nothing."

> Seth took note of his father's reaction. "I stole my dad's car, and he still didn't think there was a problem," he said. "He'd just think, 'Seth's a good kid. He cut the lawn last year.'" But Seth needed cocaine to buoy his self-esteem. In his words, "I had to use it to get the feeling of normal."

> Dining-room silver began to vanish when he was fifteen. A box of it was found to be missing, an heirloom given to his mother by a

great-aunt . . . valued at $10,000 . . . Seth had traded the wares for cocaine valued at $2,000.

I've thought about the story recently because it parallels with people's denials and suppression of addiction to polluting and depleting products like doof and activities like flying.

I have addiction stories. My stepbrother spent years in state prison because of meth and opioid use. He borrowed from nearly everyone who didn't know not to lend him money and alienated them by never paying them back. My mom is the adult child of an alcoholic father, whose habit she says continues to affect her fifty years after he died. I saw my stepfather struggle to quit smoking after a decade of unfiltered Camels. A friend took his life after becoming mired in debt and losing his home amid cocaine addiction. During the pandemic, the northwest corner of Washington Square Park—my neighborhood's backyard—became overrun with meth, crack, heroin, and fentanyl use to the point where its residents call it Crack Row. I visit Crack Row daily as part of my picking-up-litter habit, and have had several hour-long conversations with its residents.

I don't pretend to have struggled on their scales, but doof is addictive and I was addicted to it. I've felt the helplessness, hopelessness, shame, guilt, and denial, contrasted with the jolt of euphoria. I craved Doritos and Snyder's of Hanover flavored pretzel bits on the savory side; Cherry Garcia on the sweet side. I stocked them since I had my first kitchen. I'd tell myself I'd stop. I created elaborate rules limiting how much I would consume per day, yet always found myself looking at the bottom of a container I told myself not to finish.

A neighbor once left a nicotine vape pen. I stored it in a kitchen drawer. I don't smoke but was curious so tried a puff while cooking. It had a sweet apple flavor and I felt my heart beat faster, but I didn't feel better so didn't see the point. The next day I saw it so tried again, to similar results. After a few times, I noticed that though I didn't feel any joy from it, I started to expect taking a puff. I felt craving.

There is no single definition of addiction everyone agrees on, but most include consistently choosing short-term reward despite adverse consequences to yourself or others and failed attempts to stop. Addiction may result from chemicals like nicotine, alcohol, or meth, or behaviors like gambling, video games, or social media. The adverse consequences usually affect family, friends, and loved ones before the person with the addiction.

People talk about food addiction, but *food* rarely addicts. *Doof* does, the way poppy seeds don't addict but heroin does. Once someone is addicted, making what they crave available will keep them hooked, hence branding and advertising. As Coca-Cola's leader for over sixty years, Robert Woodruff implemented a strategy of putting a Coke "within an arm's reach of desire" because it worked. Now markets have flooded our world with his strategy, except more subtle and effective than he could have imagined.

According to addiction specialist Carl Erik Fisher 🎙, "Addiction is a terrifying breakdown of reason. People struggling with addiction say they want to stop, but, even with the obliterated nasal passages, scarred livers, overdoses, court cases, lost jobs, and lost families, they are confused, incredulous, and, above all, afraid. They are afraid because they cannot seem to change, despite the fact that they so often watch themselves, clear-eyed, do the very things they don't want to do."[22]

Do you like the Enlightenment values of "reason, science, humanism, and progress"? Let's see what addiction does to them.

---

22 Carl Erik Fisher, *The Urge: Our History of Addiction* (2022).

CHAPTER 6

ADDICTION OVERRIDES REASON

Tim attended one of my workshops. Before attending, he had devoted himself to promoting sustainability, which is how we met. He had also polluted and depleted heavily. He was skeptical of the workshop, seeing polluting, depleting activities as essential to modern life and stopping them as deprivation and sacrifice.

He had also gone through Alcoholics Anonymous about twenty years ago, after over a decade of out-of-control partying. Before AA, a coworker told him, "I used to drink like you. I felt like it made my life fun and exciting. Now that I don't, life is more fun and exciting."

Tim told me the man's words made no sense to him at the time, when alcohol was Tim's route to fun. "No alcohol, no fun," he felt. Since recovery, though, he has also found life more fun and exciting in ways he couldn't have imagined while drinking.

I told him I was like that coworker, telling him that whatever he feared losing when he stopped polluting and depleting, he'd find more of it, not less. Every person I've told this story to since has understood that analogy. Tim didn't see it. I stated it in different ways. I wasn't trying to persuade or convince him, just making an analogy. I realized that Tim wasn't *unable* to see it. He owned a successful business with sixty employees. He understood analogies in general. He *wouldn't* see this one.

I've had similar experiences of addicts not acknowledging their addiction. I may also have been the one not acknowledging dependencies I suppressed or denied from myself. Trying to lead an addict not ready to acknowledge their addiction to acknowledge it is like trying to catch a fish with your hand. You may drive yourself insane over their mental contortions before they acknowledge it.

Why do most addicts resist acknowledging their addiction? How do we understand what happens to reason, this deep Enlightenment value, under addiction?

Emotion, Values, and Reason

There are many ways to think about how reason works with our other mental faculties. Social psychologist and NYU professor Jonathan Haidt 🎤, in his book *The Righteous Mind*, recounts how cultures in different times and places have contrasted emotions and reason. Reason is often characterized as intelligent and as what guides emotions, which motivate action but lack intelligence. Some use similar dichotomies with different language, like limbic versus frontal cortex or lizard brain versus mammalian.

One metaphor he traced to Plato characterizes emotions as horses and reason as a charioteer. The horses create motion, but on their own don't choose direction well. The charioteer guides the direction and speed.

Jonathan modified that horse/charioteer model into a new one that I found useful and relevant. He characterizes emotion and related "automatic processes" like intuition as an elephant and characterizes reason like a rider on the elephant. He prefers "an elephant rather than a horse because elephants are so much bigger—and smarter—than horses. Automatic processes run the human mind, just as they have been running animal minds for 500 million years, so they're very good at what they do."

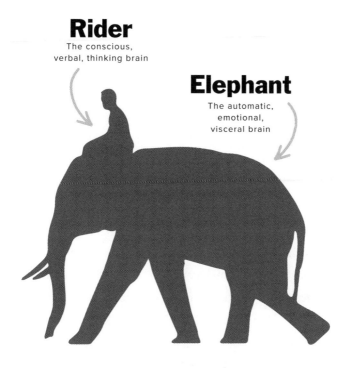

Rider
The conscious,
verbal, thinking brain

Elephant
The automatic,
emotional,
visceral brain

The Rider Evolved to Serve the Elephant

However much we may identify with the rider and value reason, the emotional and intuitive parts of our brain (the elephant) came first and the rider evolved to serve it. Intuition guides most of our behavior. Reason plays a smaller role. In Jonathan's words, "Reason serves intuition." The rider can suggest to the elephant where to go, and the elephant may take into account the rider's suggestion, but usually it does what it wants. Jonathan backs up this view with research.

The rider has an addition role: "The rider is skilled at fabricating post hoc explanations for whatever the elephant has just done and it is good at finding reasons to justify whatever the elephant wants to do next. Once human beings developed language and began to use it to gossip about each other, it became extremely valuable for elephants to carry around on their backs a full-time public relations firm."

I can also sense part of my mind rationalizing and justifying questionable things. We don't consciously tell our minds to do it. It just does.

We only experience part of our minds telling us why what we did was good, right, and natural.

Addiction and Intuition

Haidt says that we can usually trust our intuition. For most of human existence, we could safely follow it. We evolved to like the taste of healthy things like fruit and the feelings of healthy things like massages, as well as to dislike experiences of unhealthy things. No one had isolated heroin from poppy or sugar from corn.

Then we discovered the mechanism that leads to addiction—a subsystem of the emotional system. That subsystem can override the whole emotional system, which can then override our reason. Today, the things most accessible—often thrust in our faces and designed to activate that mechanism—that taste and feel best are likely to be unhealthy.

I thought to update Jonathan's model to include this mechanism for a culture where people and corporations have mastered addicting us. I thought of how we control animals like bulls, oxen, and pigs by putting rings through their noses. They are stronger than us, but those rings nullify that strength, and their intelligence. We lead them around by the nose.

Since Jonathan teaches at NYU, I run into him around campus. I proposed updating his model to include that our internal elephants can have rings put through their noses—that is, trunks—representing when we become addicted. Those rings nullify our elephants' strength and intelligence—that is, our intuition—and enable suppliers to lead us around by the nose. He agreed with the model.

Doof producers, tobacco firms, and others addict us, which commandeers our emotional system, which then commandeers our intellect to rationalize and justify behaviors those suppliers induced. Then our intellect uses every cognitive bias or trick it can to domineer the rest of our minds. Our rider goes beyond mere "full-time public relations firm" to what I call *PR-firm-on-steroids* mode.

To illustrate with an example: at the end of a meal, when the waiter asks if you want desert, you may not consciously think the following, but your intuition may be split between ordering one or not, leading your reason to guide it by considering that it may taste good in the moment but you may regret it later. If the waiter shows you a cart of deserts that you can see, smell, and almost taste, your intuition may prompt you to indulge.

By contrast, in someone addicted to cocaine, their reason has already been overridden. Offer them cocaine and their rider will likely suggest why it's good, right, and natural to indulge.

Coincidentally, the time I ran into Jonathan to share this model was in Washington Square Park, near Crack Row. It helps to look at a community of addicted people in this model.

Crack Row in the Elephant Model

In this model, Crack Row residents are people whose inner elephants have several rings in their trunks, one for each addiction. Those rings are being pulled by the relevant dealer, themselves likely pulled by their suppliers and ultimately the manufacturers. Those rings are also pulled by their fellow residents to maintain community norms.

In the model, then, Crack Row is like a network of elephants with their rings all interconnected so their behavior doesn't deviate much from each other, collectively being pulled by drug and doof manufacturers. In some ways they're like other communities. For example, people watching a football game in a stadium all cheer their teams, and people working in a company may all collaborate while being led by a CEO. In these two cases, everyone's elephants are acting in sync, but they differ from Crack Row in two big ways. First, sports fans and coworkers generally aren't motivated by addiction. They're voluntarily following shared intuitions so their elephants are collaborating, not being pulled by rings in their noses. Second, fans' and coworkers' goals usually align with their values so they lead them in directions that are healthy for themselves and others. Crack Row residents are hurting themselves and others, though their commandeered riders may nearly have convinced them that their actions are good, right, and natural.

What happens if a Crack Row resident overcomes their addiction and leaves? Like Tim unable to comprehend his coworker before AA or me now, Crack Row residents will see someone behaving differently as resisting being pulled by their nose rings. It looks painful to them, even immoral. They may describe that person in terms that imply otherness or deviance like "extreme" for violating community norms and what their riders are telling them is good, right, and normal.

Seeing what to them would cause pain, other Crack Row residents will feel they are helping those trying to leave by pulling them back in, maybe offering them drugs. People who have overcome addiction know that running into their old drinking buddies or equivalents can cause a relapse.

Overcoming Addiction Is Liberating

Addiction commandeers reason, cripples emotion, and overrides values, though few addicts notice it happening, which is why Tim couldn't or wouldn't see the parallel I described between his coworker's message and mine.

Addiction cripples us from seeing past the next hit or pang of withdrawal or from imagining a better life. We need our fix to feel normal. We prioritize ourselves over others. We don't notice the pain we cause people around us, or don't care.

Since we don't realize we're addicted, we're like if Crack Row residents cleaned their land and replanted its trees but didn't stop their addictions or supply. Most sustainability efforts I see amount to cleaning up some of our mess and planting more trees *but not ending our addictions.* Since most of our garbage no longer biodegrades, we're not even cleaning the mess. We're just moving it, which amounts to dumping it on poorer people.

Leaving Crack Row may look painful to residents and even immoral, but it is *liberating.* It creates mental freedom.

When I avoid doof and flying, to mainstream people, my behavior may look painful, like deprivation and sacrifice, and even immoral, but it is *liberating.* It creates mental freedom. It's relaxing. I'm removing rings others attached to my elephant's trunk.

What the Spodek Method Does

The Spodek Method shows that you can remove your rings of addictions based on polluting and depleting. That's the mindset shift. Removing the rings takes effort and time. That's the continual improvement. It liberates us from being controlled by that mechanism. It enables us to expect that the more rings we remove, the more we liberate ourselves, so we can learn to look forward to what we would otherwise unconsciously avoid. Until we start, though, our riders in PR-firm-on-steroids mode will try to convince us otherwise.

People addicted to doof, or even those who just consider doof normal, will see me as struggling, and extreme, but I'm not. I got rid of my elephant's rings so I'm free, mentally and physically. The more rings I remove, the more I want to remove, no willpower necessary.

This metaphor helped me see another motivation beyond what I realized after Frances's memorial. I'm also climbing those stairs and experimenting to liberate myself and help others liberate themselves. It bears repeating: **I'm not acting on something extreme or unique to me. I'm acting on universal human emotions and values. I'm not helping "the environment" in the abstract; I am helping people here and now.**

My Strategy in the Metaphor

You read my mission on page 3: to change American and global culture to where people expect acting more sustainably will bring not deprivation or sacrifice but liberation and freedom. You know my core tactic is the Spodek Method. Haidt's vocabulary enabled me to state my strategy to use that tactic to accomplish that mission.

In short, my strategy is to enable people to liberate themselves by showing them that they can remove those rings, make them aware that they'll be glad they did and wish they had earlier, then show them how. Even if they stop there, they will have improved their lives, but most feel motivated to liberate others. So the bottom-up part of my strategy is also to train people how to liberate others and train people to train people and so on—to create a self-sustaining, accelerating movement.

In this metaphor of riders on elephants with nose rings, influential people pull on more people's rings than others. The top-down part of my strategy is to lead renowned people to remove their rings in open, genuine, authentic ways.

Even if you can't change anyone else, liberating yourself improves your life. You can also learn to lead others with it. As we'll see, even that outcome vastly understates the value and potential of this burgeoning movement because we are suffering from more than the addictions I've described so far.

Are Polluting and Depleting Behaviors Addiction?

We've heard of people flying private jets to climate conferences. They say they're helping and probably believe they are, but it's hard not to view their beliefs with skepticism. I see them like people on Crack Row, only polluting and depleting millions of times more. As best I can tell, they believe their behavior is good, right, and normal, at least they say so, though they wriggle like Tim, so I suspect their riders are in heavy PR-firm-on-steroids mode.

For a while, I described behaviors enabled by polluting and depleting like buying large SUVs and flying as addiction. To clarify: flying may pollute and deplete, but here I'm pointing out that polluting and depleting enable flying the way alcohol enables partying. Such behaviors fit the criteria of addiction since they led people to choose behaviors for short-term reward despite adverse consequences. Such behaviors are like drinking where you got the party and someone else on the other side of the world got cirrhosis. Those activities hurt the actors eventually too; addiction generally hurts others before the addict.

But many people buying SUVs and strawberries in December don't see the connection between their behaviors and the environmental problems making headlines. They know the ocean is filling with plastic, which harms people (and wildlife), but they won't see *their* disposable coffee cups as contributing. The class of literature on the environment from ecomodernists, techno-optimists, Pinker, and their peers promoting efficiency, decoupling, and substitution fits Tim's pattern. They're promoting practices that accelerate our environmental problems, but they don't acknowledge what's happening.

All these people's views seemed to come from riders beyond just being in the PR mode Haidt described. They seemed to be in PR-firm-on-steroids mode. Were they addicted?

CHAPTER 7

THE PSYCHOLOGY OF CORRUPTION

We don't feel like we're addicted. Why not?

Beyond polluting and depleting, we insist that other people should change, not us. I have yet to meet a person who insists on taking responsibility for hurting others with their polluting and depleting behavior. Everyone knows about refugees, birth defects, and so on, but we all have stories about why we aren't the ones who should change, that our plane was going to fly anyway, that our plastic will not reach the ocean, or some story that says that others should change but not us. We're one of the good guys. Our actions feel good, right, and normal.

Why We Support Polluting and Depleting Activities Against Our Deepest Values

We see the headlines of people (and wildlife) suffering. We may have reasons for polluting, like visiting distant relatives, but we can't deny that we contribute to that suffering. What happens when I believe I'm a good person but try to reconcile doing something I believe is wrong?

Most of all, I can't escape my internal conflict. In principle, I could stop doing the thing, but if my culture does it, then, like Jefferson and Washington, I may not overcome the hurdles to stop. I could change my values, but when the values I'm breaking include the Golden Rule, stewardship, and common decency, changing them will hurt me more than help me. Putting it simply: if I do something that kills people and I don't want to kill people, I have to stop doing it.

If I don't stop, then I have to convince myself that what I believe is wrong is right. Trying to convince myself that what I believe is wrong is right forces me to deny and suppress what I plainly see.

Two death spirals result. The first is internal: hurting people leads to internal contradiction leading to hurting them more. This cycle leads to suppressing and denying the internal contradiction leading to trying to convince ourselves even more that we are good. The pattern leads from internal conflict to suppression and denial, to lower self-awareness, and ultimately to increasing harm and internal misery.

INTERNAL CONFLICT DEATH SPIRAL

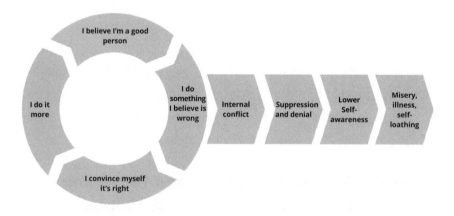

The second death spiral is external and leads us to resent the people we hurt when we want to believe they aren't suffering. We also come to resent the people who tell us about this suffering. Seeing their suffering causes us to face our guilt, shame, and other results of corruption. We might think: "I didn't feel guilty. Then I learned about someone suffering. Now I feel guilty. The person who told me about that suffering caused the guilt, or the person suffering did," instead of facing that our consciences are causing us to feel guilty.

EXTERNAL CONFLICT DEATH SPIRAL

I call the result of insisting we're one of the good guys while feeling internal conflict, misery, and resentment toward others *pollutionism*. It promotes us polluting more. Pollutionist beliefs include:

- "What I do doesn't matter."
- "Only governments and corporations can make a difference on the scale we need. *They* caused the problem. *They* should fix it."
- "Any change I could effect would make my life difficult and, divided by eight billion, wouldn't make a dent in the rest of the world. I care about the environment, but I have to balance."
- "Individual action just doesn't make sense. BP made up this idea to distract us from their responsibility."
- "We need to create technologies to solve our problems. That's progress."
- "We need to progress to help all those living in poverty around the world. It's not fair to deny them the progress we benefit from."
- "If we don't grow, we'll return to the Stone Age."
- "If you disagree, you're indulging in ghoulish fantasies of a depopulated planet with 'Nazi-like comparisons of human beings to vermin, pathogens, and cancer.'"
- "More people means more geniuses to solve our problems."

When Enough People Support Polluting and Depleting, Polluting and Depleting Become Our Culture

A third death spiral occurs at the cultural level. When enough people say something like pollutionism is good, right, and normal, however much we believe polluting and depleting are wrong, pollutionism becomes a cultural norm and we stop questioning it.

Our beliefs don't end with accepting pollutionism. That acceptance decreases internal conflict that could create civil unrest so it serves cultural institutions. Authorities support "proven" results, creating what I call *scientific pollutionism* to institutionalize pollutionist beliefs and relieve people with consciences wracked with guilt and shame of having to question themselves.

CULTURAL CONFLICT DEATH SPIRAL

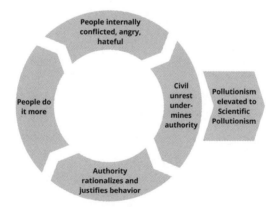

Scientific pollutionism includes beliefs like:

- "We can decouple the economy from polluting."
- "We can dematerialize, substitute, and create efficiencies to where polluting and depleting become non-issues."
- "Economies need to grow."

- "Economic growth has lifted more people out of poverty than any other system, so we must keep growing to help the poorest people."
- "Solar and wind are 'clean,' 'green,' or 'renewable.'"
- Techno-optimism.
- Ecomodernism.

Future generations will dismiss scientific pollutionism, but it appeals to us. It enables us to believe we are helping those we hurt. It tells us, independent of the consequences of our actions to others: "you're a good person."

Scientific pollutionist "evidence" and "proofs" don't need to be right. They only have to ease our consciences enough to sleep at night.

Pollutionism Is PR-Firm-on-Steroids Mode

Does the pattern of "proving" oneself right look familiar?

Consistently acting against our values leads our inner elephants to put our riders beyond the ordinary PR mode Haidt described into PR-firm-on-steroids mode too. The deeper the values and more we transgress them, the deeper the rider goes into that mode.

Going into PR-firm-on-steroid mode happens physically in our brains and bodies. The mechanism activating that mode in many polluting, depleting behaviors may be different than with drugs, though closer to behaviors like gambling, shopping, or playing video games. Industries design mechanisms to make those activities addictive.

It's easy to see the denial in others who are addicted, but as Tim showed, it's harder to see our own, especially when our community supports us and our PR-firm-on-steroids rider tells us after every questionable choice that we're good, right, and normal (and people who tell us otherwise are the opposite). We don't consider ourselves addicted to flying, doof, or SUVs, because our PR-firm-on-steroid mode riders call them necessities, rationalizing and justifying with: "I want to visit family flying-distance away," "I'm hungry and there's only doof around," "I live in Phoenix and the heat is intolerable," "We have a baby and can't avoid disposable diapers and toys," and so on. "What else can we do?" we ask.

It may only be a matter of time before psychiatric associations recognize behaviors enabled by polluting and depleting as addictions, but we can apply what we've learned from understanding and treating addiction to help people who want to pollute and deplete less.

Not Hopeless

I'm not pointing out our addiction to make us feel bad, but to show the problem and its solution. Education, legislation, technology, and market regulations help only so much and often exacerbate the problem. Facts, statistics, instruction, and punitive laws rarely stop addiction. More effective is listening, understanding, supporting, providing role models, and changing environments, though every case is unique. People don't rationally choose to become addicted, so trying to reason them out of it becomes 4C-bludgeoning.

We can get our inner riders out of PR-firm-on-steroids mode. I've hosted guests on my podcast who have hit rock bottom of addiction then emerged and became inspired leaders. Many former addicts find meaning and purpose in helping others overcome their addictions. Their shared experiences often make them most able to help.

Still, how much have we made our world Crack Row? Obviously not completely, but we'd be irrational and in denial not to see many losses of freedom and reason that didn't exist more than a century ago and keep growing. What happens when we live increasingly without reason? Let's examine our world.

CHAPTER 8

WHAT WIDESPREAD ADDICTION DOES TO US

I worked with the founder of a prominent nonprofit whose mission was to decrease plastic pollution. He worked with plastic producers and the UN on its plastic treaty. He told me that his daughter loved strawberries and the only way to get them for her year-round was to buy them shipped from California in plastic. As much as he wanted to end plastic pollution, he loved his daughter and wanted to make her happy. He kept buying the plastic and funding the problem he made his life's mission to stop. **But you can't stop the global heroin trade if you're worried about your own supply.**

We're addicted to things like strawberries in December, comfort, convenience, disposable diapers, and the ability to fly anywhere on a whim. Addiction doesn't feel like you're a bad person. We usually experience the opposite: our inner riders make it feel good, right, and normal. It feels like warmth, love, and support. The prospect of stopping doesn't feel like liberation. It feels like denying your daughter the strawberries she loves. It feels like risking returning to the Stone Age.

Another person I worked with, Eugene, was dedicated to sustainability before we met. He works to find ways to live more sustainably and help others to too, but he has a problem. He's American, his wife is Japanese, they live in Hawaii, his family lives in California, and hers lives in Japan.

They live a challenge common in our culture: flying-distance families. Many flying-distance families claim they can't help flying. If we define addiction as a behavior we can't stop despite adverse consequences, flying qualifies.

I hear the strain in Eugene's conscience when he talks about lowering his impact by using wooden toothbrushes even though he hasn't reduced his flying, which pollutes and depletes millions of times more. The issue is not others' judgment, but his own internal contradiction between *his* behavior and *his* values. It's easy to procrastinate. If you fly, it's hard to think of the full effects beyond your benefit, but we can't fly without making people refugees by how we extract the minerals and fuel from their land, causing global warming, oil spills, extinctions, and so on.

Crack Row's residents' addictions feel to them normal, comfortable, and warm, even loving. So do ours. They "need" their fix to feel normal. So do we. Can you imagine not flying? Many people describe not flying as impossible. We've flipped our perception of reality and impossibility. No human flew before 1903, when even top physicists considered it impossible. If you fly, you probably feel like nearly everyone does, but you're in a minority of a few percent of the global population.

Despair, Helplessness, and Hopelessness

"You guys, look! Sarah and Dave brought heroin!"

These words were spoken cheerily by the host of an upscale suburban dinner party, as someone might announce Sarah and Dave had brought homemade pie. The scene was in a romantic comedy, *Seeking a Friend for the End of the World*, where everyone knows an asteroid will hit the earth that is big enough to destroy all human life. People are responding in many ways. Some smoke the Cuban cigars and drink the fine liquor they'd been saving. Others get depressed. In the scene of the upscale party, one character had been telling another about his sexual escapades with partners seeking physical pleasure. When Sarah and Dave arrived with the heroin, he called out, "Bucket list! Bucket list!" and went over to inject.

They had chosen to live "eat, drink, and be merry for tomorrow we may die." That sentiment means that if the future can't be better than today, then we can't beat momentary pleasure and we might as well go with it. Why care about meaning, purpose, or the future?

With all the predictions of environmental destruction, people who can't imagine a better tomorrow settle on "eat, drink, and be merry for tomorrow we may die." It means giving up and procrastinating. Those claiming "faith that the next generation will solve these problems" have given up on their own generation. Those promoting colonizing Mars have given up on earth. Those promoting technology that doesn't exist yet have given up on themselves.

It's tempting to give up. I did before my experiments. At first I wanted to prove that giving up was my best option. But to conclude no solution exists because I can't think of one is a failure of imagination. To conclude living sustainably can't work is a failure of leadership.

I talk to people on Crack Row because I've learned that they aren't an aberration of modern society but a more acute version of it. They are our future if we don't change.

When people interact directly, we tend to practice values embodied in *Do Unto Others As You Would Have Them Do Unto You* (the Golden Rule), *Leave It Better Than You Found It* (stewardship), and *Live and Let Live* (common decency). These values exist in many cultures, likely most. Many of us practice them all the time, but when our behavior is mediated through the environment, our culture abandons them. If we want to fly across the country to go to a meeting or party and come back a few hours later, we only consider if we have the time and money for it.

We have developed an implicit understanding, as if we say to each other, "I'll say what I do doesn't matter. You say what you do doesn't matter. Neither of us will question each other and if someone else does, we'll point at governments and corporations and say it's their fault."

We don't ask, "How will flying affect others?" We get annoyed at anyone who asks because we know the answer: it will hurt them. We feel:

"If I can't do anything about it, how does making me feel guilty change anything?" and "I just want to relax. I don't want to have to think about everyone else every time I do anything."

We've replaced the Golden Rule, stewardship, and common decency in interactions mediated through the environment with giving up.

What Addiction Takes From Us

You may think, "So what if I'm addicted if I don't feel bad from it?" If addiction isn't that bad and ending it seems to make life worse, why stop?

That's how Carl Witonsky felt as he insisted his son was fine, even as he stole the car, TV, and heirloom silver. If you asked him, he'd insist their lives were wonderful. Their home had a pool and tennis court. But any outside observer would say Carl would be better off acknowledging the situation. Suppressing, denying, rationalizing, and justifying adverse effects causes them to recur. Addiction robbed Seth, Carl, and Pecki of love for each other and they didn't see it.

What are we missing without seeing it?

You Tell Me What You Fear Losing and I'll Tell You What You'll Gain

Addictions involve a reward that feels good. Gamblers feel like winners. Meth users feel energy. Nicotine users feel calm. Social media users feel connected. But outside a brief jolt, addiction deprives you of that feeling in the rest of life. Gamblers feel like winners but they're losers—literally: they lose money. Meth users feel more energy briefly but have less overall. Social media users feel connected, but become isolated.

The following graph, from addiction specialist Dr. Anna Lembke, 🎤 author of *Dopamine Nation*, illustrates the effect. Dopamine is a brain chemical linked to feelings of reward. Addictions tend to result from activities creating dopamine. The more we engage in them, the more our bodies develop tolerance, meaning the highs become smaller and shorter, the lows become deeper and longer, and our base state stays lower. We feel we need to do the drug, activity, or doof just to feel normal. The result of ending an addiction: *you tell me what you fear losing and I'll tell you what you will gain.*

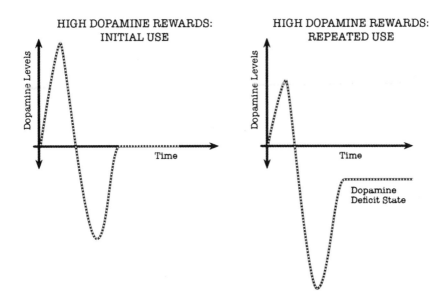

Do you think you need to fly for family or work? Do you think we need to pollute to reduce poverty? Stop polluting and depleting and you will experience *more* connection to family, *more* control over your career, *more* nature, *more* adventure, *more* leisure, *more* to look forward to, *more* cultural exchange, *more* comfort, and *more* convenience. You will help the poor the most. You'll have to pass through withdrawal, but it's temporary.

People considering giving up heroin will find it hard to imagine regular life matching its predictable euphoric jolt, but like the construction worker who gave me twenty dollars, they know it's possible.

If I talk about stopping using disposable cups, people tell me they don't have time to sit to drink coffee. But people who commit to avoiding disposable cups in favor of mugs end up having more time, not less. They find themselves more productive, not less.

As people increasingly get coffee to go, they seem more hurried but get less done. Same with ordering more takeout and binge-watching more TV, among other busy activities. When you see the pattern, it's like meth users thinking they're full of energy while overall losing it and doing meaningless busywork, distracting them from meaningful engagement.

Only By Understanding the Problem Can We Solve It

As with doof, I point out our addictions to other substances or behaviors not to make you feel bad, but to clarify the problem and make it easier to acknowledge in ourselves. People rarely overcome or even acknowledge their addictions from being told facts, numbers, or instruction.

Understanding the nature of doof and addiction helped me to see that I was out of control, not just nibbling a few extra pretzels. Seeing industries benefiting from my dependence helped me see that dependence. It wasn't personal weakness in me but industries with vast resources engineering products to take advantage of the human emotional system. They're piercing my inner elephant's trunk with more rings and driving my rider to promote what they're selling. I felt as much disgust at doof manufacturers for promoting their products as food as I did at tobacco companies marketing their products to children. That emotional disgust toward their business practices expanded to include physical disgust for their products, which I reinforce by picking up litter daily since it's mostly doof wrappers and Amazon.com packaging.

Finding role models helped me face and overcome unwanted dependencies, as did books and videos on addiction. It helped me to see addiction not as something abstract but something existing in all of us.

Efficiency Increases Addiction

We believe that efficiency will decrease waste. With things that addict, we achieve the opposite. When morphine was refined from opium—that is, made more efficient—people expected it to lower opium addiction. Addiction increased. Then heroin was made yet more efficient. Medical experts expected it to decrease addiction but it increased again. It's obvious with clearly addictive things:

poppy → opium → morphine → heroin → fentanyl

Consider these product evolutions. People felt increasing efficiencies would decrease problems.

tobacco → cigars → hand-rolled cigarettes → mass-produced cigarettes → vape pens

wine → distilled liquor → mass-produced liquor

walk → horse → train → car → plane

conversation and gossip → newspaper → online media → social media (→ artificial intelligence?)

Instead we have a world of fentanyl, vape pens, cheap liquor, flying-distance families, and social media. We're more efficient than ever, polluting and depleting more than ever too. Everyone feels good about each development until seeing the results.

Not seeing the pattern has led us to create a culture of busyness, meaningless activity, isolation, self-interest, inability to reflect, joylessness, myopia, self-righteousness, defensiveness, and not caring about the effects on others. We've developed a new cultural morality based on efficiency conflicting with our innate morality based on the Golden Rule, stewardship, and common decency. We're torn up inside too.

CHAPTER 9

IT'S NOT FAIR: WE DIDN'T CAUSE OUR ADDICTION, BUT ONLY WE CAN OVERCOME IT

Our situation seems the height of unfairness. The Sackler family and its company Purdue Pharma created OxyContin, an addictive opioid. Knowing its addictive capability, they promoted it while downplaying its risks. One of the business world's most prestigious organizations, the consulting firm McKinsey, advised them not to reflect or constrain themselves but to "turbocharge" sales. They addicted millions, profited from it, and used those profits to shield themselves from justice. Those addicted, their families, and their communities are suffering and dying. Yet even if the Sacklers and McKinsey wanted to help, only an addicted person can end their addiction.

This pattern is repeating throughout the economy. Addiction is profitable. The products sell themselves.

If you start smoking, you know you're taking a risk. The box warns you that nicotine is addictive. Supermarkets don't sell illegal drugs. By contrast, with doof and things polluting and depleting bring, we didn't choose to take a risk. I grew up learning that flying was good so didn't know I risked hurting others or myself. Doof was marketed as food. Disposability was marketed as good, as was social media.

We couldn't have anticipated the outcomes would become adverse.

Confounding the problem, the adverse effects hurt others (like alcohol where we get the party and someone else gets cirrhosis), making it harder to motivate stopping.

Only an addicted person can face their withdrawal. If they don't, they can't end their suffering. If they do, they still risk relapse. Those who benefited from addicting them can't do it for them.

My stepbrother told me his doctor prescribed him OxyContin for an injury, telling him it was safe. Next thing he knew, he couldn't function without it. After the prescription ended he could buy it illicitly. Next thing he knew, he needed to sell the drugs illegally to afford buying and ended up in prison. It's not fair that after others help themselves in hurting you, only you can overcome your addiction.

Today, we all face this situation to some degree. Exxon hid research that its products would overheat the globe, but we have to endure withdrawal if we want to overcome addiction to polluting and depleting behaviors, not its shareholders. We built cities like Phoenix in the desert, highways that create car-dependent suburbs, and flying-distance families. You and I didn't create the system driving these results. We didn't ask for it. We didn't choose to risk getting addicted. We're like people who were told OxyContin wouldn't cause addiction by those profiting from it. Now the Sacklers and their helpers keep the money and power to corrupt the justice system to protect themselves while our communities disintegrate, people die, and addiction rates rise.

No one intended to create a culture causing us to believe family, work, and food require polluting, depleting, and hurting others. We didn't choose to be born into it.

Others profit from our addiction. They profit from improving technologies and techniques to addict you more, and faster. It's tempting to think that because governments and organizations made the stuff that polluted their bodies and minds did it, *they* should have to act. They should act in many ways, but to overcome our addictions, they can't. Until we overcome our addictions, we'll keep funding them.

It's not fair that nearly everyone around you is addicted and acts in ways to keep you addicted as if they are helping you. It's not fair that it's easier not to try to overcome your addiction and easy to relapse if you do. It's not fair that marketing to feed your addictions exists everywhere.

It's not fair that if you overcome your addictions that hurt others, you may never see the benefits since people on the other side of the world may be the ones who feel relief. It's not fair that the prospect of ending your addiction feels like losing support, warmth, protection, even love—sometimes feeling like losing the only source of such things in what feels like an otherwise cold, cruel world.

For a community to overcome addiction requires ending supply *and* demand by helping everyone overcome their individual addiction. If even a few people in the community remain addicted and maintain their supply, they can cause everyone to relapse.

Even If It's Unfair, We Can Still Do It

Still, we need not feel hopeless or helpless. People have changed culture at the global level before. It may not be easy or fast, but we can do it. We can learn from history to change faster and more effectively.

My mindset shift helped. Now I feel disgust knowing that every dollar I spent on my addiction funded suffering. I funded kicking people off their land, causing disease and birth defects, putting children in mines, and so on. I still do, but over 90 percent less. I'm working to decrease more and to change the systems that make it hard to stop. That knowledge about what you fund may help you when you learn it, repel you, or may help you later, after your mindset shift. However trapped I felt in activities that pollute and deplete, only I could take responsibility for my actions.

Many people have overcome addictions as deep as the ones I'm describing. We can use them as role models. We can overcome our addictions, even to flying and doof. We'll be glad we did and wish we started earlier. The longer we go, the easier it becomes. We will find our past addictions disgusting, repulsive, repugnant, dreadful, and nauseating. We will find reward in helping others overcome their addictions too.

PART 3

THE DAWN OF UNSUSTAINABILITY: HOW WE GOT HERE AND WHAT IT DID TO US

CHAPTER 10

HOW A CULTURE OF CRUELTY EMERGED AND ENDURES

The more I experimented with living more sustainably, the more I found myself exiting a culture I hadn't questioned. I could see the cruelty that my culture caused to the people (and wildlife) on the receiving end of the pollution and depletion. Again, the point isn't judgment but internal conflict. We're human. Few of us want to be cruel, so how did a culture creating cruelty emerge?

Recall, I thought my entire sustainability journey would last one week of avoiding packaged food so I could say I tried and justify giving up. Instead, the more I explored living more sustainably, the more I researched what people already living without what I was giving up did. Increasingly I looked to indigenous cultures, which surprised me. I'd learned to see them as stuck in the Stone Age. Wasn't thirty old age for them? Didn't they die from infections without antibiotics?

The indigenous culture I learned of first was the San in southern Africa, from the book *Affluence Without Abundance*, by anthropologist James Suzman 🎙, who had lived with them for decades. I invited him to the podcast, along with people who had lived with the Hadza in Tanzania, Kogi in Colombia, Tsimane in the Amazon, and Matsés in Peru. I attended events hosted by local Native Americans.

As I learned to live more sustainably and learned more of these cultures, I came to see they made it look easy and preferable. I also saw they shared properties with each other but apart from us. I couldn't tell if there was an "us" and "them," but I saw hints. I saw that my old belief that they were living in the Stone Age was ignorance on my part. So was my fear that the Stone Age was the only alternative to modernity.

Kevin Didn't Make the Fence.
The Fence Made Kevin.

I'm about to describe processes of hierarchy forming, evolving into empires, and spreading around the world. When I write about people long ago or far away, it's tempting to consider ourselves different—"How could they do those things? *I* never would!"—but they're as human as we are. Here's a story you may find familiar to help bridge that divide.

I volunteer with a group that delivers overstock food to local community fridges that anyone can take from. I mainly deliver to the Chelsea Fridge, which was started a few years ago, coincidentally by a former student. A church on 15th Street had a fenced-off area where it stored garbage. She coordinated with the church to install a fridge and cabinet there instead. When I unplugged my apartment, volunteering a few evenings a week came to replace screen time.

Kevin Fucillo 🎤 also volunteers and spends the most time around the Chelsea Fridge of all the volunteers. He delivers food, organizes, helps other volunteers unload, finds new stores to donate, and helps maintain order. If other guests take too much, he reminds them others are hungry too. Last year I began hearing about a guy Kevin coordinated with who delivered food by the truckload, ten times more than the cartloads most of us delivered. The deliveries came at different times than I delivered, so I only heard about these giant deliveries.

Recently my schedule changed and I started seeing these deliveries. He texts Kevin when he'll arrive, usually the same time each week, but not always, so guests only learn through Kevin.

By the time I saw one of these deliveries, Kevin had developed ways to keep order, enlisting other regulars he knows well to put the most desired stuff behind the fence. He established a set of rules on dispensing it, so far unwritten.

Unloading can take thirty minutes. Kevin makes sure no one takes until everything is unloaded. Guests help and can see everything available and figure out what they want that others might take first. Some guests position and grab more aggressively than others. Guests generally show order and mutual respect, but regularly a guest will knock boxes of food over or step on food to get what they want. When guests break rules, he may announce that they may be banned.

If you ask Kevin, I think he'd say he's doing what's necessary to keep order and protect the food from damage. Conflict, if noisy, could lead neighbors to complain, which could end the deliveries. He's not taking more than his share, so he's not trying to benefit materially. Still, guests congregate near the fence, trying to access the more desired stuff, and Kevin dispenses it his way, often making them wait. He and his friends may give people they know some in advance.

I'm not saying he's doing anything wrong. On the contrary, he's fulfilling a role without which no one might get any food. Maybe it's just me, but he seems to have more energy when he's in charge than other times. He didn't start by saying, "How can I create a hierarchy?" He took responsibility when no one else did. If someone else did, they'd probably do something similar.

The Chelsea Fridge is not unique. Anyone who has worked in an office has seen this pattern. Even in the Chelsea Fridge community, a parallel situation is occurring on the online forum where volunteers coordinate. Sabrina moderates it and coordinates with the church so she has similar status which she has used to block volunteers from shifts. I suspect she sees the situation like Kevin: she didn't try to create hierarchy or status, she just took responsibility and is trying to optimize food delivery. Kevin's and Sabrina's domains don't often conflict, but sometimes do, leading to friction.

You've likely seen hierarchies form and people attain status in your communities. If you've been subordinated, maybe you've felt frustrated at being forced to go through them. If you've acquired status and exerted authority and people complained, maybe you felt misunderstood, since you likely acquired that status by taking responsibility to help the team.

Anthropologists study situations like this. They might describe the pattern as a dominance hierarchy forming with subcultures within it. They ask why a hierarchy forms the way it does, why some hierarchies subjugate others ("dominance hierarchies") while others don't, why people develop the subcultures they do, how big a hierarchy can grow, what happens when hierarchies conflict, and so on. In the rest of the book, when I write about dominance hierarchies forming and evolving into empires with cultures exhibiting subjugation and cruelty, if you wonder how people would just let it happen, think of your version of the Chelsea Fridge situation.

For example, when I say subcultures form among those with status that differ from subcultures among those without status, it sounds clinical, but I can feel myself adopt subcultures based on my status. For example, Sabrina has an authoritarian style, which I don't like, but accept, despite feeling internally conflicted at accepting something I don't like. Some things I tell myself to handle that internal conflict: it's no big deal, let it slide; I could talk to her and suggest she lighten her grip; I could propose people rotate between roles, but it feels like too much work; maybe I should organize other volunteers to create systems to maintain order with less hierarchy. Have you thought similarly?

When I've had status and exerted authority, I've told myself other stories. Have you ever though things like, "This job isn't easy. If they don't like it, they can do it, but they aren't and it has to be done. Why do they make it difficult for me? Can't they tell that if someone didn't do it, everything would fall apart? How can I get them to act for the team instead of for themselves?"

It's tempting to think something about Sabrina led her to play that role as she does, like she's hungry for power, but we've seen similar situations happen with other people. Learning the patterns shows us that something about the conditions are causing the situation more than her personality traits. In the rest of this book, when learning about emperors and imperial subjects behaving cruelly, it will be tempting to say: "They do that because they're just cruel. They're different than me. I wouldn't do such things." If so, I hope you'll consider hierarchies that have enveloped you, roles you've played, and stories you've told yourself to justify participating in ways you wouldn't otherwise. Our hierarchies likely didn't grow that large or subjugate people that much, but we likely didn't limit those outcomes. Material conditions probably did. If they hadn't, how much might we have been swept along?

Hierarchy and Status Didn't Make the Surplus and Fence. The Surplus and Fence Made Hierarchy and Status.

The biggest lesson to learn from the Chelsea Fridge is that Kevin and Sabrina didn't create the conditions that made a hierarchy form. The conditions that made a hierarchy form created Kevin and Sabrina's status at the top and corresponding dominant behavior, as well as mine below.

We're individuals that act on our own, but I learned from anthropology that regarding the patterns of hierarchy I describe below, people are doing what you and I would likely do in their place; maybe not exactly, but over enough time and involving enough people, cultures will converge.

We Think Our Culture Is Best, Yet People Choose Against It in Head-to-Head Competition. Why?

I started learning anthropology because it inspired me to unplug my apartment on the particular day that I did. That morning I read *Tribe: On Homecoming and Belonging* by anthropologist Sebastian Junger 🎤, which I read a couple months after *The Dawn of Everything* by anthropologist David Graeber and archaeologist David Wengrow.

Both books recounted how during colonial times in North America, people who experienced Native American culture and colonial culture preferred Native American, contrary to European and colonialist expectations. Both books quoted Benjamin Franklin: "When an Indian child has been brought up among us, taught our language and habituated to our customs, if he goes to see his relations and make one Indian ramble with them, there is no persuading him ever to return." The pattern included colonists who had experienced Indian life: "Tho' ransomed by their friends, and treated with all imaginable tenderness to prevail with them to stay among the English, yet in a short time they become disgusted with our manner of life . . . and take the first good opportunity of escaping again into the woods." Disgust is a powerful emotion. I had been feeling it more and more from picking up litter, seeing people suppressing their awareness of others' suffering resulting from their lifestyle choices, and so on.

They quoted John Hector St. John de Crèvecoeur, a European who lived in North America, in 1782: "Thousands of Europeans are Indians and we have no examples of even one of those Aborigines having from choice become European. There must be in their social bond something singularly captivating and far superior to anything to be boasted of among us."

Among people who lived in both cultures—that is, in head-to-head competition—people chose the one I'd learned to see as primitive or uncivilized. The pattern seemed backward from what I learned in school. Wasn't the European culture more advanced? Shouldn't everyone have preferred it?

Columnist David Brooks wrote in "The Great Affluence Fallacy" in 2016 in the *New York Times:*

> In 18th-century America . . . the settlers from Europe noticed something: No Indians were defecting to join colonial society, but many whites were defecting to live in the Native American one. This struck them as strange. Colonial society was richer and more advanced. And yet people were voting with their feet the other way . . . . During the wars with the Indians, many European settlers were taken prisoner and held within Indian tribes. After a while, they had plenty of chances to escape and return, and yet they did not. In fact, when they were "rescued," they fled and hid

from their rescuers. Sometimes the Indians tried to forcibly return the colonials in a prisoner swap, and still the colonials refused to go. In one case, the Shawanese Indians were compelled to tie up some European women in order to ship them back. After they were returned, the women escaped the colonial towns and ran back to the Indians.

Were my feelings of liberation and freedom like these women's? Once I tasted it, I wanted more—freedom from harming innocent people, from craving, from being controlled, from being led to buy things, and most of all, from the torment of the death spirals commandeering my intellect into PR-firm-on-steroids mode. Since people around me were like the colonists who couldn't see what people preferred to their culture—freedom, equality, and democracy—they thought I was giving up something valuable. I wasn't. I was taking rings out of my inner elephant's trunk.

It wasn't just Native Americans and colonists. *Tribe* described cases throughout history and around the world of people who chose what I had thought of as less advanced, more primitive cultures. That head-to-head preference belied the story that I belonged to a culture that was improving itself and the world. I believed that centuries ago people lived as serfs, eking out miserable existences, and that further back, during the Stone Age, they barely survived, lucky to find food at all before being eaten themselves. It turns out that the pattern these books described was old news to anthropologists. Their data had long ago disproved "linear" models of cultural evolution.

Junger described the pattern in other situations, especially where civilization broke down, such as after natural disasters. Status went away. Everyone had to help each other. He reported that civilians who had been trapped in war situations, like the Siege of Sarajevo or London Blitz, found morale improved, even as they were shot at and bombed. Many later reminisced fondly about the community and equality they experienced then. Soldiers returning from tours of duty often reenlist. Junger himself was embedded with American soldiers at a remote outpost in

Afghanistan where they were shot at nearly daily. When they returned to the US, many chose to reenlist. He made a documentary about it called *Restrepo*, which was nominated for an Oscar. People choose how to live for many reasons, but that people in perhaps the most prosperous society in history preferred a war zone where they knew they would be shot at seems to clearly indicate their values.

We who haven't lived immersed in indigenous cultures or survived extended disasters can't fully compare such situations to ours, but among those who have, in head-to-head comparisons, *once they experienced alternatives, people often chose the alternative, to the point of being shot at.* What made the alternatives so preferable?

Tribe partially answered: relative to their cultures, ours offered less freedom, equality, mutual support, family ties, and abundance, resulting in higher rates of suicide, depression, addiction, isolation, overwork, scarcity, and more.

What I learned from *Tribe* would have been interesting on its own. Combined with the results from my experiments, it forced me to see my culture anew. I was still living in Manhattan. I lived far from sustainably, but I was taking steps toward it. The beliefs I learned growing up told me each step should worsen my life, yet I was experiencing the opposite. *Tribe* showed that people have been discovering this pattern for centuries.

How did my culture look to people from indigenous cultures? *The Dawn of Everything* recounted how European culture looked to a Native American Wendat named Kondiaronk, who toured Europe. He remarked, "Do you seriously imagine that I would be happy to live like one of the inhabitants of Paris? To take two hours every morning just to put on my shirt and make up? To bow and scrape before every obnoxious galoot I meet on the street who happens to have been born with an inheritance? Do you actually imagine I could carry a purse full of coins and not immediately hand them over to people who are hungry."[23]

---

23   David Graeber and David Wengrow, *The Dawn of Everything* (2021).

When asked what would happen if monarchy and status were done away with, he responded, "A leveling equality would take place among you, as it now does among the Wendat and yes, for the first thirty years after the banishing of self-interest no doubt you would indeed see a certain desolation as those who are only qualified to eat, drink, sleep, and take pleasure would languish and die, but their progeny would be fit for our way of living. Over and over I have set forth the qualities that we Wendat believe ought to define humanity: wisdom, reason, equity, etc."

I wondered how other cultures viewed us. I asked Alan Ereira 🎤 , who lived among the Kogi in Colombia and produced two BBC documentaries on them. Without pause, he responded, "vicious," adding that they saw us as cruel and working too hard. When among them, he felt cared for. He said the idea of stepping over a homeless person would be inconceivable to them. In London, he felt he had to turn his senses off from the sensory assault of our culture.

Dawn of Everything showed many ways people lived that dismantled the model I had learned that my culture was more advanced and theirs was primitive. Humans have thrived in countless cultures with countless structures. We *haven't* followed a linear path of progress.

Myths Busted

I still couldn't see what, if anything, distinguished "my culture" from "alternatives," but the more I learned, the more I saw what I had considered obvious truths crumble.

I thought that for most of human history thirty was old age. It turns out that "the average modal age of adult death for hunter-gatherers is 72 with a range of 68–78 years. This range appears to be the closest functional equivalent of an 'adaptive' human life span."[24] That is, my belief that thirty was old age was wrong for 250,000 years of our ancestral past.

---

24 Hillard Kaplan and Michael Gurven 🎤, "Longevity Among Hunter-Gatherers: A Cross-Cultural Examination," *Population and Development Review* 33, no. 2 (June 2007): 321–365.

In the last century, we merely restored what was likely normal before agriculture and industrialization. Moreover, the greatest causes for our increased longevity were sanitation and hygiene, which don't require polluting and depleting.

I "knew" we worked fewer hours on less grueling tasks than a century or two ago. I figured medieval serfs worked longer and harder, but it turns out they worked fewer hours per day, fewer days per year.[25] Hunter-gatherers appear to work fewer hours, with healthier diets and greater food security.[26]

Kondiaronk had no formal education, yet outperformed Europe's intellectual elite. It turns out he wasn't special among the Wendat. Rather, they practiced what I would call effective leadership skills: listening, conflict resolution, rhetoric, influence, and such. They learned these skills through regular civic interaction.

Speaking of cultures being more egalitarian, I thought that modern society was overcoming racism, implying that it was worse in the past, more so the farther back we'd look. Instead, Oxford-educated Trinidadian historian Eric Williams wrote, "A racial twist has thereby been given to what is basically an economic phenomenon. *Slavery was not born of racism: rather, racism was the consequence of slavery.*"[27] [my emphasis]

Recently, historian Ibram X. Kendi agreed. He wrote, "I had been taught that racist ideas cause racist policies. That ignorance and hate cause racist ideas. That the root problem of racism is ignorance and hate. But that gets the chain of events exactly wrong. The root problem," he continued, "has always been the self-interest of racist power. Powerful economic, political, and cultural self-interest . . . has been behind racist policies. Powerful and brilliant intellectuals . . . then produced racist ideas to justify the racist policies."

---

25 Juliet B. Schor, *The Overworked American: The Unexpected Decline of Leisure* (1991).

26 James Suzman, *Affluence Without Abundance: The Disappearing World of the Bushmen.* (2017).

27 Eric Williams, *Capitalism and Slavery* (1944).

Several big questions dogged me. The biggest was

> If people preferred Native American culture in head-to-head competition, why did colonial culture win?

Next:

> Why does my culture see living sustainably as undesirable or impossible, even privileged?

> Why does our culture want to bring our polluting, depleting behavior to others, like cell phones and hydroelectric dams? Why do many people in other cultures reject them?

Those questions seek to understand our situation. Looking forward:

> Why does sustainability not mean returning to the Stone Age?

> How can we get the best of both—the advantages of other cultures and our own?

And regarding my sustainability leadership results:

> Why do workshop participants, after their mindset shifts and continual improvements, describe experiencing *more* freedom when using *less* technology and innovation? What was technology bringing if not freedom?

> Why do they start seeing many results of what our culture calls "progress" as burdensome? Why do they start describing Costco as "repugnant"? Why do they choose to avoid flying?

I'll give you highlights of the answers, as best I understand them so far, and point to where we can learn more. The experience of living more sustainably post-mindset shift will reveal the answers more than writing can. To clarify, I didn't only learn good things about "them" and problems with "us." I also learned about great things about my culture, which also partly answers why the colonists won.

Defining and Distinguishing Cultures

I had been describing my culture as "our polluting, depleting culture" and others as "indigenous," but the labels felt inaccurate for my purposes. *Indigenous* means the culture was there first, which may be a valuable distinction in some contexts, but not here. I didn't know if cultures I was learning from were there first. Even if the Hadza had lived as they do for fifty thousand years, how do I know they didn't displace other people? How do I know if the people Europeans displaced from Manhattan didn't displace others already there? Should I presume the people who settled places first are most worth emulating?

How do I decide if indigenous cultures are worth learning from? Native Americans who open casinos on their territory may be indigenous, but they seem assimilated and promoting addiction. Were the Khmer of the Khmer Rouge army that killed over a million people in Cambodia in the 1970s indigenous? Should we support an indigenous Nigerian who takes control of an oil well and partners with Exxon for being indigenous? Should we commit infanticide or genital mutilation because indigenous cultures do?

I saw what mattered for sustainability in distinguishing cultures was *behavior*. The relevant core properties of what I saw as my culture result from its behaving unsustainably. I identified the distinguishing behaviors as polluting, addicting, imperialism (more on imperialism to come), and depleting, which I simplify with the acronym PAID, as in "our PAID culture."

I could call the cultures that people preferred "non-PAID," but since they existed first, I didn't want to define them for what they weren't. The opposite of polluting and depleting seemed to be *sustainable*. The opposite of addicted seems to be *free*. The opposite of imperialism seems *abundant*, as I'll clarify below.

I started calling the cultures I was learning from "Sustainable Free Abundant." We Americans enjoy freedom, equality, and abundance compared to many other PAID cultures, but we lack the freedom, equality, and abundance of Sustainable Free Abundant cultures like the San, Hadza, Tsimane, Kogi, and Matsés, and situations Junger described. I was beginning to sense my culture's lack of freedom, equality, and abundance just by avoiding packaged food and flying. Unplugging the apartment increased these things, along with teamwork, meaning, purpose, empathy, and compassion.

Context to How PAID Culture Began

Our ancestors evolved into humans about 250,000 years ago. As far as we know, they lived as hunter-gatherers from then until about twelve thousand years ago.

In that time, they spread to six continents. If Malthusian collapse were inevitable, they had more than enough time to have outgrown their food supply. They were resilient. They survived changes in climate, variations in environments from tropical to arctic, and more. Combined with what we know of today's hunter-gatherer cultures which most resemble our ancestors', cultures that endured likely practiced what we would call stewardship, not straining their environment for their needs—hence Suzman's title, *Affluence Without Abundance*. They had cultural guardrails to maintain resilience and stewardship—not cutting down too many trees, not hunting the last animal, and leaving something for nature.

They also seem to have practiced equality more than we do. A common guardrail in Sustainable Free Abundant cultures: when members try to brag or create hierarchy, the rest of the group pokes fun at them. For example, if a hunter keeps claiming to bring back the best catch, others will say it looks like skin and bones. A community may kick out a hunter who continues, no matter their skills.

Anthropologists call the period of highly varying climate the *Pleistocene*, which started two-and-a-half million years ago and ended when the climate warmed and stabilized about twelve thousand years ago. They call the current period the *Holocene*. Human activity didn't stabilize the climate, but that stabilization changed us. In particular, in a few river valleys, agriculture became advantageous, suggesting that agriculture was impossible during the Pleistocene but inevitable during the Holocene.[28, 29]

Agriculture revolutionized more than food production. Two changes in particular are relevant to sustainability and our culture today. First, agriculture created surpluses. Second, agriculture developed in places that were bounded, so could be protected: "Each had a resource that was amenable to intensive exploitation . . . Each also was in some way bounded geographically and environmentally."[30] That is, geography enabled control over food supplies. For the first time in human existence, the Holocene created material conditions like the Chelsea Fridge's surpluses and fence.

Surpluses in a bounded place are the conditions for *dominance hierarchies* to form. Dominance hierarchies in human culture take away freedom from those at the bottom and, as we'll see, force those at the top into behaviors too. They open the door to chiefdoms, then cities, states, and nations. If unchecked, they also lead to tyranny, authoritarianism, dictatorship, autocracy, despotism, and other non-democratic structures.

Understanding dominance hierarchies helps us understand how they affected humanity and led to our culture today.

---

28 John L. Brooke, *Climate Change and the Course of Global History: A Rough Journey* (2014), and Jared Diamond, *Guns, Germs, and Steel* (1997).

29 Peter J. Richerson et al., "Was Agriculture Impossible during the Pleistocene but Mandatory during the Holocene? A Climate Change Hypothesis," *American Antiquity* 66 (2001), 387–411.

30 Brooke, *Climate Change and the Course of Global History*.

CHAPTER 11

DOMINANCE HIERARCHIES, PROTO-TYRANNY, AND TYRANNY

Dominance hierarchies exist in many social species and emerged long before humans evolved. They are one type of social hierarchy. You can have hierarchy without dominance, for example someone learning a craft as an apprentice to a master. There, the term master comes from mastery of the craft, not being a master of other people. Apprentices may take instruction from the master, but if they can leave if they want then they are not *dominated*. Sports leagues may have hierarchies including world-class professional, minor-league, amateur, and recreational, but if professional players order amateur players around, they can walk away.

In a dominance hierarchy, people higher in the hierarchy (higher *status*) can impose their will on people lower than them. Dominance hierarchies aren't necessarily good or bad. They can create order and stability, which people may like when the hierarchy helps everyone. Kevin and Sabrina created order at the Chelsea Fridge. When I sail, for example, I value the dominance hierarchy under a skilled, experienced skipper, especially in difficult weather when the skipper knows more about sailing than anyone else on board. Surgeons in operating rooms don't take votes on what scalpel to use; they tell the nurse which one to hand them.

A few cultures today, including the Hadza and San, seem *egalitarian*—that is, without dominance, even of the most common types: by sex (usually patriarchy) or age (gerontocracy). Many anthropologists believe most of our human ancestors for 250,000 years lived in egalitarian bands. In such groups, if I try to order you to do something, you can walk away or poke fun at me. If I want to influence you, I have to talk with you. Kondiaronk was more skilled at leadership than most Europeans because he grew up with little dominance. Among the Wendat, interacting with others required everyone to learn social skills from childhood. Europeans largely lived in dominance hierarchies then—tyranny, in the words of the Declaration of Independence—so few of them learned the social and emotional skills of leadership.

When high-status people benefit from polluting, depleting, addiction, and imperialism, they can use their status—that is, their control over necessary resources—to protect themselves from the harms that PAID activities cause. Doing so reinforces the hierarchy, potentially leading to tyranny.

To remove tyranny, it helps to know how it started. Anthropologists and historians trace our current global hierarchy through various incarnations to the dawn of agriculture.

What causes dominance hierarchies?

My years experimenting living more sustainably made me feel like the people Junger and Brooks described—the women who had to be tied up to be returned to "civilization" and then ran away and the veterans who preferred reenlisting and facing enemy fire to mainstream American culture. I was curious: how did PAID culture form and defeat Sustainable Free Abundant cultures? Nobody chooses to work hard so someone lazing around can take the fruits of their labor.

Seeing what caused dominance hierarchies and tyranny made everything fall into place for me.

What causes a dominance hierarchy to form? It's tempting to say physical strength, since males dominate females in many cultures, but matriarchies exist, and men with status in patriarchies aren't usually the strongest. Some people suggest that some people's thirst for money or power, or traits like cruelty drive them to create dominance hierarchies. Others suggest that things like political power cause them to form, but then what is political power?

A dominance hierarchy can form when two conditions are met

1. There is a necessary resource that can be controlled
2. There is no alternative to it

When sailing at sea, for example, the skipper controls access to my safe return and I have no other option. Once we dock, that hierarchy goes away. These conditions are why I emphasized that agricultural surpluses began in river valleys: "Each also was in some way *bounded geographically and environmentally.*" The rivers and stable climate provided the first condition to form a dominance hierarchy. Being bound provided the second.

The story doesn't end there, because stored surplus prompts competition. Even if a community with a controllable resource wants equality and devises ways to maintain it, if other communities can take the surplus, the community has to protect it.

Competitive Strategies: How and Why Dominance Hierarchies Grow

Understanding when and how hierarchies formed reminded me of lessons in business school about competition: if a firm has "barriers to entry" and "customer captivity"—business lingo meaning roughly that it meets both conditions for dominance hierarchies—it can dominate its market.

Bruce Greenwald's course "The Economics of Strategic Behavior" was among Columbia Business School's most popular when I attended the school. He led the school's renowned Value Investing Program. His peer and friend, Warren Buffet, graduated from the school too. In a book

Greenwald co-wrote, *Competition Demystified*, I reread decades after graduating how he described dominance hierarchies in business terminology:

> If there are barriers, then it is difficult for new firms to enter the market or for existing companies to expand, which is basically the same thing. Essentially there are only two possibilities. Either the existing firms within the market are protected by barriers to entry (or to expansion), or they are not. No other feature of the competitive landscape has as much influence on a company's success as where it stands in regard to these barriers.

He described how this situation led to strategies for responding to competition. He calls those with status "elephants" and everyone else "ants." "For an elephant operating within the barriers, life is sweet and returns are high." That is, elephants dominate. But he didn't stop analyzing at a sweet life with high returns. Those returns come from customers, who might not like paying so much, so they may try to work around a company flexing its ability to dominate. Competitors, regulators, and others may want to end its domination.

Greenwald continues, "Competitive advantages still have to be managed. Complacency can be fatal, as can ignoring or misunderstanding the sources of one's strength. An elephant's first priority is to sustain what it has, which requires that it recognize the sources and the limits of its competitive advantages." He lists four top strategies to manage those competitive advantages:

1. Reinforce, protect, and extend existing advantages
2. Distinguish where it can profitably grow or might be undermined
3. Focus on policies that maximize profit
4. Spot competitive threats that require strong countermeasures

When Greenwald says "complacency can be fatal," in a business context he means a business can go bankrupt. When you dominate people as a warlord or emperor, fatal can mean you being deposed and killed. It's why governments offer witness protection for people who testify against organized crime bosses. If organized crime bosses lose their status, they can die, so killing witnesses makes sense to them. Likewise, if North Korea's rulers relinquished power, many of their victims and their families would want revenge, so complacency in implementing Greenwald's strategies would be life-and-death fatal. As historian Timothy Earle stated, "A close examination of history indicates that only a coercive theory can account for the rise of the state. Force, and not enlightened self-interest, is the mechanism by which political evolution has led, step by step, from autonomous villages to the state."[31]

It's tempting to attribute to warlords and emperors traits like desiring power, but calling them vicious, cruel, or inhuman leaves unexplained why kingdoms, empires, and hierarchies didn't form before agriculture. People didn't change ten thousand years ago, the climate did. Material conditions didn't *allow* dominance hierarchies to form and grow. They *forced* them to. They didn't *allow* cruel people to rise to the top, they *evoked* cruelty in those with status and *selected* cruel people and elevated them. Then their inner riders led them to believe what they did was good, right, and natural (and everyone else's inner riders to believe their subjection was good, right, and natural too).

Just as racism didn't cause slavery but slavery caused racism; cruel and violent people didn't make unsustainable tyranny. Unsustainable tyranny evoked people's cruelty and violence. These cultures result from *material conditions* combined with strategies that if you were complacent in implementing you risked dying. When we take those conditions for granted, it's tempting to see the people implementing the strategies as great conquering emperors (or villainous tyrants), but taking them into account, cultural evolution looks more like water flowing downhill. The more you know the terrain, the more you can predict where it will flow.

---

31   Timothy Earle, *How Chiefs Come to Power: The Political Economy in Prehistory* (1997).

You can change where water flows more effectively by changing the terrain than by pushing the water. The strategies led to many of PAID culture's values, which feel normal to us even when they contribute to cruelty, like nine million people dying per year from air polluted by our purchases and other activities, as reported by *The Lancet's* Commission on Pollution and Health in 2022.

Every step from egalitarianism to PAID likely seemed as optimal to people in the moment as it did to Kevin and Sabrina. Can you blame someone for:

- Discovering plants could be domesticated?
- Then accumulating a food surplus?
- Then defending the surplus?
- Then developing a militia and culture to support it (as well as arts, music, and so on)?
- Then expanding to supply the home market?
- All the while, creating rationalizations and justifications for expanding?

As for ants, Greenwald advises "exit gracefully," meaning for a business without barriers to entry to exit the market. Businesses may exit markets or declare bankruptcy without any human suffering. Actual humans can't always peacefully exit dominance hierarchies. We may have to accept subjugation, as I did with Sabrina's authoritarian style.

As an exercise to the reader, when you see lack of freedom, equality, and democracy, look for the necessary resource being controlled. That control is the source of political power. In agriculture the controllable resources are arable land and surplus food. What was being controlled during the Atlantic Slave Trade?[32] In the antebellum American South?[33] Today?[34]

Understanding the source of hierarchy in controlling necessary resources without alternatives can help us identify how some can subjugate others. The terms "global north" and "global south" attribute status to compass position, which is unrelated. For example, there is a region in Louisiana between Baton Rouge and New Orleans so polluted from oil refineries that it has become known as Cancer Alley. People there live in the global north but lack the status of corporate executives running fossil fuel companies a bike ride away who may have among the highest status in the world. Meanwhile, the dictator of a resource-rich nation south of the equator who controls access to it may live in the global south but have high status. Instead of identifying people's statuses by distracting irrelevancies, we could identify them by their level of control over necessary resources.

PAID culture is sustained by energy sources that can be controlled, which today include fossil fuel, nuclear, wind, solar, hydroelectric, and geothermal energies. Status derives from how much access one has to them. If you live a lifestyle that uses a lot of them, you have high status. People who can't control it and have to accept its harms have low status. High status people subjugate low status people to make their phones, cars, and so on—often indirectly, but mainly in how they spend their money. We may have high status in some contexts and low status in

---

32 Partial answer: Weapons, ships, and other technology; a justice system; arable land.

33 Partial answer: Along with weapons and a justice system, you might think arable land, since it produced cotton. It turns out that until the eighteenth century, the land produced little profitable cotton because processing took too much labor. Unsustainable agriculture had exhausted the land. Ironically and tragically, a technology designed to decrease labor—Eli Whitney's cotton gin—made a strain of cotton profitable over a much larger area.

34 Fossil fuel extraction is controllable by whoever owns the well or mine, likewise for uranium and other nuclear fuels. As long as we depend on them for energy, fertilizer, steel, plastic, microchips, and so on, we reinforce the hierarchies. Solar, wind, and hydroelectric energy look dispersed, but manufacturing, installing, operating, and disposing of the equipment to convert them to electricity require fossil fuels (nuclear reactors do too), so using them reinforces the hierarchies.

others. Cancer Alley residents may still drive and fly. Rich people's children can still get asthma.

This last description makes people in PAID culture sound cruel and indifferent. We don't feel cruel and indifferent. Why not?

Status Forms Complementary Cultures

Greenwald's strategies lead people with different status to behave and think differently—that is, they create different cultures. High-status people adopt beliefs and cultures that rationalize and justify their status and cruelty—beliefs like the divine right of kings, trickle-down economics, that a rising tide lifts all boats, and racism.

When low-status people see their chance of escaping subjugation not worth the risk, they will form cultures rationalizing and justifying inaction and acceptance. They may believe in things like reward in the afterlife, "the meek shall inherit the earth," or that "we shall overcome." They may try to convince themselves that "individual action doesn't matter," "the plane was going to fly anyway," and "only government and corporations make a difference."

Until agriculture and unsustainability, cultures that survived lived sustainably. They didn't have to conflict over resources, so they developed cultures that considered stewardship, resilience, equality, and freedom good, right, and natural. As unsustainable cultures grow, they become elephants, making sustainable cultures into ants, whom they crush, rationalizing that they are helping them.

Both groups will feel their beliefs as true, and acting on them as good, right, and natural. If high-status people value helping others, they will feel that everything they do helps others, even if cruel, like slavery, or polluting, depleting, and addicting. Low-status people will convince themselves that inaction is the best they can do. They aren't misguided or gullible. They're responding to their conditions. Switch someone's status and they'll adopt the beliefs of the new status.

CHAPTER 12

THE RESULTS OF UNSUSTAINABILITY

Other characteristics of these cultures cause other behaviors. A surplus in food means not enough of some other necessities if a culture grew its population to match the food supply. Then it will run out of other things like metals, minerals, and labor. Agriculture often exhausts the land. These characteristics mean that a culture requires things beyond its immediate environment or ability to trade with other cultures with enough surplus—that is, its people are living unsustainably.

Unsustainability requires growth. Unsustainable elephants (of Greenwald's type) interacted with other elephants too. What does that interaction look like and lead to?

When Elephants Have Their Way

Areas like the Fertile Crescent and river valleys around the world saw unsustainable cultures grow, conflict with each other, and become more aggressive. As they kept growing, they increasingly conflicted with other aggressive, growing, unsustainable elephants—that is, other dominance hierarchies. I recommend watching several videos online that wonderfully illustrate this effect.[35]

---

35   Joshua Spodek, "Visualizations of Empires Growing and Competing," JoshuaSpodek.com, April 18, 2024.

Their conflicts led them to develop weapons and political structures to defeat competitors for resources. Then they adopted beliefs, stories, and cultures to rationalize and justify taking over. When two elephants competed, both could survive, one could defeat the other, or both could die. Some once-dominant cultures may have died, but others grew to empires dominating beyond river valleys to large swaths of East Asia, the Indian subcontinent, and the Mediterranean.

Meanwhile, these unsustainable cultures kept needing resources beyond their borders. They had few options: to collapse, become sustainable, or keep expanding. Countless unsustainable cultures have collapsed. Some unsustainable cultures have become sustainable, but rarely, since they would fear becoming ants that elephants would stomp on.

The third option—to keep expanding—means plundering the abundances of Sustainable Free Abundant cultures. Success with this plunder would restore abundance to the unsustainable culture, but only temporarily if it stays unsustainable. Then it will exhaust that abundance, resulting in a bigger unsustainable culture. If it keeps plundering, it will accelerate the cycle.

This practice has a name: *imperialism*. When imperialists take another culture's land, it becomes *colonialism*. When they force them to work, it becomes *slavery*. They all result from unsustainability, which requires growth, and Greenwald's strategies to maintain dominance. In Haidt's metaphor, dominance hierarchies forming and evolving into slavery changed people's internal elephants (keeping Haidt's metaphorical elephants distinct from Greenwald's). Before unsustainability, people were egalitarian, interacting voluntarily, following their intuitions. Over time, more and more rings were put in their internal elephants' trunks, depriving them of more and more freedom, coercing them more and more. PAID culture is a giant mesh of people's elephants tugging each other by their rings toward replenishing their resources depleted from living unsustainably. Meanwhile, everyone's PR-firm-on-steroids riders did more than rationalize and justify their actions. They trumpeted "growth is good" and "we're civilizing barbarians," not seeing that they were abandoning stewardship.

Before the dawn of unsustainability, stewardship was easy. Everyone's behavior affected everyone they knew. If you hurt someone, you knew it. Nothing was addictive. Everyone's elephants were aligned and no one's reason was crippled. Today eight billion people are being pulled by their rings away from stewardship. We've had ten thousand years of people saying growth is good because growth always appeared to help (not counting the many cultures that collapsed). The concept of toxic pollution that didn't biodegrade barely existed, if at all.

Properties and Values of Imperialist Cultures

Greenwald's strategies led to elephants—that is, imperialist cultures—to evolve common properties and values. We may attribute the properties to people, but the cultural evolution was like water flowing downhill. Let's consider some cultural properties resulting from imperialism.

When imperialist cultures increase in size and complexity, they develop arts, music, architecture, a complex economy, and other elements of culture. People with authority in these PAID cultures will develop beliefs and stories that their cultures are good, right, and natural—in particular, that their culture is better than those of the ants they destroy. They will tend to lose connection with nature and with each other. The bigger and more experienced their cultures grow, the better at imperialism they become. They will become unable to see that the people and cultures they destroy prefer their own ways in head-to-head competition. Beliefs emerge to rationalize and justify, such as the divine right of kings, that others are barbarians or savages, and racism.

Proto-tyrannies evolved into tyrannies. The general trend was toward larger empires with more arts and culture, but also less freedom, social mobility, and equality. The result: more intensive agriculture, slavery, less resilience, less stewardship, fewer but greater collapses, larger militaries, more cruelty, more disease, loss of knowledge and connection to nature, more competition between cultures, exhausting soil, depleting forests, depleting biodiversity, and so on.

Imperialist cultures would value control of land and energy over less tangible measures Kondiaronk valued, like quality of relationships, quality of life, resilience, freedom, equality, and democracy. Under imperialism, the property someone controls indicates a person's status. The incentive to work people as much as possible for as little pay as possible grew.

Imperialist culture teaches its youth to acquire control over others more than to relate with others, self-awareness, or other social and emotional skills. It reduces politics from groups deciding things together to people with status implementing Greenwald's four strategies.

Imperialist civilizations were on track to exhaust the soil and deplete other resources several centuries ago. Instead, three unexpected things happened that changed everything. To understand the scope of the impact of these unexpected happenings, imagine yourself in, say, medieval Europe, running into limits on arable land for food and wood for energy, heat, and housing.

Imagine discovering

1. An infinite amount of arable land
2. An infinite source of energy that didn't require chopping down trees

These discoveries could have removed the conditions for a dominance hierarchy. Freedom, equality, and democracy could have emerged, but Europe had lived so long under tyranny that few knew how to practice these concepts. But imagine also discovering:

3. Cultures that were experienced and skilled in freedom, equality, and democracy you could learn from as role models

Wow! Now you could recreate freedom, equality, and democracy.

How the Enlightenment and PAID Culture Formed

These three discoveries happened. Europeans discovered:

1. What looked to them like infinite land in 1492
2. Energy in coal that looked infinite too, ironically prompted by depleting their forests

And soon after:

3. Cultures like Kondiaronk's practicing freedom, equality, and democracy

European cultures that had competed for centuries began collaborating. Land that appeared infinite meant energy and food looked infinite too. They began living as if these things were infinite and their cultures evolved to match. Their arts, literature, and philosophy advanced.

But their practice of stewardship of land decreased since they could exhaust soils and consume nonrenewable resources without regard to the future. They could just move on. Their practice of stewardship of human relationships plummeted with indentured servitude and slavery.

Europeans and American colonists began to learn new world practices and beliefs, which didn't derive from strict dominance hierarchies. "Especially with the Spanish conquest of the Americas," recounts *Dawn of Everything*, "suddenly, a few of the more powerful European kingdoms found themselves in control of vast stretches of the globe, and European intellectuals found themselves exposed, not only to the civilizations of China and India but to a whole plethora of previously unimagined social, scientific and political ideas. The ultimate result of this flood of new ideas came to be known as the 'Enlightenment.'"

These three discoveries alone didn't cause the Enlightenment, but they contributed significantly: "Benjamin Franklin, Thomas Jefferson,

James Madison, James Wilson, John Adams, Thomas Paine, and other founding fathers were acquainted with Indian peoples, tribal governments, and indigenous theories of governance," wrote historian Robert Miller.[36] Indigenous wisdom contributed to the Enlightenment and American democracy. Miller continued,

> Indian nations and peoples impacted the formation of the current US government, the Constitution that created it, and several specific provisions in that document . . . The hundreds of years of interactions between native nations and English and American colonies, states, leaders, and the United States founding fathers shaped the political thinking of both sides and even influenced the development, drafting, and ratification of the US Constitution.

Slavery led to growth of markets for products like sugar and tea that were unnecessary, but which the home markets couldn't stop buying. I considered them close enough to addicted to say that while these three discoveries may have brought the Enlightenment, they also led to pollution, addiction, imperialism, and depletion. They caused European culture to become full-fledged PAID culture.

These discoveries explain the trend to today's peace that Pinker wrote about. The Enlightenment values of "reason, science, humanism, and progress" didn't reemerge from humans becoming more innovative or more insightful. They emerged from material conditions no longer promoting a dominance hierarchy.

Pinker ascribed the Enlightenment's development mainly to human innovation like the printing press and increased trade, but humans didn't store the energy in coal, create the Americas, or create cultures like Kondiaronk's. We *found* them. In this regard, the onset of the Enlightenment resembled the onset of the Agricultural Revolution resulting

---

36 Robert J. Miller, "American Indian Constitutions and Their Influence on the United States Constitution," *Proceedings of the American Philosophical Society* 159, no. 1 (March 2015) 32–56.

from the climate changes of the Holocene period: we didn't cause it, but we took credit for it. If our cultural change resulted from conditions we didn't cause, it could change again if our environment changed again. We might have a false sense of security over our ability to solve problems.

Problems Despite the Enlightenment

PAID culture came with downsides.

Entitlement: Assuming infinite land, energy, and food meant that material limits could make once-productive practices counterproductive, like exhausting soils and abandoning them for new land or working slaves to death. Two continents were big enough and the fossil fuels (and later uranium, solar, wind, and water power) were plentiful enough that their presumption that these things were infinite remained untested for a couple centuries, but the land and energy sources they found *weren't* infinite.

Inability to resolve conflict: For centuries, European nations fought over resources. After discovering land and energy that seemed infinite, they increasingly expanded outside Europe. They didn't learn to resolve conflicts. Eric Williams saw this inability to resolve conflicts: "When, in 1492, Columbus, representing the Spanish monarchy, discovered the new world, he set in train the long and bitter international rivalry over colonial possessions for which, after four and a half centuries, no solution has yet been found."[37]

Lost meaning and isolation: Using coal enabled us to do more things. Today, we value spending less time washing clothes, but everything done *for* us deprives us of the sense of accomplishment *from* doing it ourselves. We lose meaning and purpose. It's tempting to say "people used to spend all day washing their clothes. The washing machine liberated us," but that's our PR-firm-on-steroids talking. Sustainable Free Abundant cultures work less than we do.

Questionable values: Finance and economics grew in importance over

---

37 Williams, *Capitalism and Slavery*

citizenship and social and emotional skills. Transactions grew in importance over relationships.

Beyond valuing growth, PAID culture *requires* growth: We today are so used to valuing growth we take it for granted as a proxy for freedom, equality, democracy, jobs, and other things we value. Unsustainability causes the need, as it causes imperialism, colonialism, and slavery. "We must expand or perish," said Georgia Senator and slaveholder Robert Toombs in 1860 (soon to become a secessionist, stating, "We want no negro equality, no negro citizenship; we want no negro race to degrade our own.") He continued, "We are constrained by an inexorable necessity to accept expansion or extermination."[38] If we end slavery but keep living unsustainably, we'll continue the imperialism and colonialism that led to slavery.

Pollution: Before the mid-nineteenth century few could have anticipated it. It had barely existed then. What pollution did exist, as far as they could tell, like smoke from burning wood, could be resolved by "the solution to pollution is dilution," as people in the twentieth century said. Forest fires existed before humans and their smoke dissipated. How could they imagine carcinogens, endocrine disruptors, plastic in our arteries (and placentas), or the greenhouse effect? Earth's reservoirs were big enough that it would take centuries to fill to the point that new pollution hurts people immediately, so they kept polluting more.

Degradation: For millennia, farmers exhausted soils and armies salted fields, but crops eventually grew back. How could they have imagined topsoil shrinking and disappearing or forests as large as continents turning to desert? They didn't understand extinction or keystone species.

They blinded themselves to how their actions affected others. Expanding cultures across the globe with supply chains just as long, they distanced themselves from people affected by their decisions. A landlord in England could instruct people at a sugar plantation in the Caribbean to

---

38 "The South Rises Again—and Again, and Again," *New York Times,* January 2011.

increase production and not know the deadly results. Consumers could buy sugar and not know the cruelty they were funding.

But the two biggest problems would happen even with infinite land, food, and energy.

First, the Enlightenment restored *some* values—reason, science, humanism, and progress, as Pinker noted—but not *all* the values that enabled cultures to live sustainably in the long term. They didn't restore stewardship—that is, living sustainably either with nature or other people. We'll see below how deeply Enlightenment figures and US founding fathers valued stewardship and intended to maintain it, but they didn't in practice. Why should they? Believing in infinite land told them they could exhaust all the land they wanted to for profit and move on (likewise with people they enslaved). They isolated themselves from nature and from each other by rationalizing and justifying what they did as good, right, and natural.

Consider free trade as business school taught me. When everyone is free to choose, more trade means more mutual benefit. Then profit is a proxy for good. Believing their trade was free trade, they (and we) used more trade as a proxy for more good, but much of their biggest trading was coerced—slaves, sugar, and cotton. They promoted more coercion and cruelty, mistakenly congratulating themselves for helping the people they hurt. Their PR-firm-on-steroids rider told them that their slaves "feel an affection for their master, his wife and children" or that Africans mated with orangutans.

They called cultures that resisted "Stone Aged" or "uncivilized." Once they made those cultures dependent, they called them "developing." Their PR-firm-on-steroids riders blinded them from seeing that other cultures preferred to live as they had for thousands of years.

Huge surpluses in food, land, and energy meant greater imbalances with other resources, especially labor, metals, and minerals. These relative shortages, combined with isolation *increased* imperialism, colonialism, and slavery, even amid the Enlightenment's values. Augmented by technology, these patterns ultimately led to the greatest slave power in history in the American South.

We Believe Coercive Markets Are Free

For a market to be free, people must be able to choose if they want to participate in an interaction. If so, it can create mutually beneficial results. A free market allocates resources to people best able to solve problems. By contrast, a coercive market allocates resources to people who can best coerce.

Consider a market based on slavery. That market is not free but coerced. In a slave-based market, people with many slaves accumulated great wealth. If you convinced yourself of racist beliefs, you'd believe your market was free when it wasn't and say that your wealth showed you helped the world.

Consider a market based on polluting and depleting. They destroy others' life, liberty, and property without their consent. When I pollute, I coerce people who cannot consent. Therefore, that market is not free. It is coerced. It doesn't allocate resources to people best able to solve problems. It allocates resources to people who coerce the most. When the people we coerce are remote and we convince ourselves of pollutionist beliefs, we wrongly conclude that wealth comes from improving the world. Some people have become wealthy that way. Many others, though, have become wealthy from coercion.

Twentieth Century Problems: Addiction and Marketing

Speaking of coercion, the twentieth-century introduced two new problems, perhaps two sides of a coin: deliberate addiction and marketing based on *wants* instead of *needs*. For most of my life I adopted PAID culture's values and beliefs uncritically. I strove to excel at them.

Addiction wasn't new. Vedic literature wrote over two thousand years ago of gambling addiction, but only after the Enlightenment came to value growth without stewardship did culture begin to promote it.

Addictive products sell themselves! Cracker Jacks were introduced in 1896, with the slogan, "The More You Eat The More You Want," making them one of the first products designed to addict that was promoted as food. The following year saw heroin synthesized, which Bayer marketed as a "non-addictive morphine substitute."

Legal addictions included gambling, doof, tobacco, shopping, and OxyContin. Many illegal ones started as legal, including cocaine, heroin, amphetamines, and tobacco marketed to children.

The other problem that came with the twentieth century was the shift in marketing from needs to wants, led by Edward Bernays.[39] Needs can be met. Once you have enough food, water, and a few other necessities, you can move on to things like hobbies, family, arts, sports, and civic participation. Wants are malleable and can persist forever. Marketers connected Virginia Slim cigarettes to female empowerment, Ford F-150s to masculinity, diamonds to marriage, and so on to create unlimited demand.

Some results: Instead of singing, we learned to buy music recordings to listen to passively. Instead of telling stories around a fire, we watch TV. Instead of playing sports, we watch them on TV. Instead of traveling by actively moving, we are transported around the world. Believing we have infinite energy, we build tools to work for us. They provide comfort, which we take for granted, at the expense of appreciation, purposeful effort, accomplishment, and meaning. We isolate ourselves where we used to collaborate. We become passive consumers instead of active citizens.

Recall that the Industrial Revolution's first wave employed some of the greatest minds of its time in mechanical engineering and later waves in civil, chemical, and computer engineering. Today, many of the world's largest, most profitable companies hire psychological engineers

---

39 Amy Westervelt 🎙, *Drilled*.

to create products, advertising, and more to manipulate our reward systems to cause addiction and craving. Enlightenment values without stewardship promote these patterns.

As soon as wants-based marketers get a foothold in your psyche, Greenwald's strategies lead them to increase your dependence and find more wants. They lead us to believe that we need them for the most important parts of life. In increasing realms of our lives, they've learned to control our emotional systems better than we can. Machines doing physical work for us made our lives easier but made us physically weaker and dependent. Increasingly, they think and choose for us and "there is no statistical record of any other period in US history when people have spent more time on their own . . . Swapping touches for screen taps, America's kids are experiencing a more solitary, and melancholy, childhood than we've ever seen," with similar patterns in adults.[40]

We Pay For Our Misery

We in PAID culture have convinced ourselves that ours is the best and others want to be like us. We call other cultures "developing." Thus, for centuries we couldn't understand why people from other cultures resisted assimilation, to the point of preferring being shot at. Here are some key differences I see:

---

40 Derek Thompson, "Why Americans Suddenly Stopped Hanging Out," *Atlantic*, February 14, 2024.

PEOPLE IN SUSTAINABLE FREE ABUNDANT CULTURES	PEOPLE IN PAID CULTURES
love more	buy more
are more free	are more addicted and controlled
are more secure	are more insecure
support each other more	are more needy
are more resilient	are more fragile
become mature while young	remain infantile into adulthood
are surrounded by nature's beauty	pave over and fear nature
embrace purposeful effort	procrastinate in fear of losing comfort
are more fit	are more obese
enjoy delicious food	drench their foods in salt, sugar, and fat
are helpful	are helpless

Addiction and dependence create dominance hierarchies as effectively as invading with less violence and risk, so imperialists and colonialists have long used them. They also enable invaders to frame themselves to their home markets as helping the cultures they plunder. People in the home market feel that they are helping those whose cultures they destroy. Today, we bring them cell phones, doof, casinos, and other addictions that cement that dependence faster and more securely.

Is it obvious why those women had to be tied up to be returned to PAID culture from a Sustainable Free Abundant one and why they escaped when they could?

Is it obvious why I'm so enthusiastic about restoring sustainability and stewardship? We can help today's equivalents of those colonists become free and not tyrants.

We in the Home Market Drive the Process. We Are Powerful.

When I think of imperialism, I tend to think of Roman legions with SPQR flags or European slave ships off the coast of West Africa. Those soldiers and sailors are the visible part of imperialism and may commit the most violence, but they didn't start the process, nor do they drive it. They're risking their lives far from home and family.

Why risk their lives? Because their government pays for or compels them to profit from pillaging, often collaborating with corporations. But even these governments and corporations don't buy their own goods. People in the imperialist culture's home market do. The home market drives imperialism, first by exhausting its surplus, then by buying what was taken from others. Europeans created sugar plantations in the Caribbean because they bought so much sugar, otherwise the plantations would close. If no one bought slave-produced cotton, the American South would not have grown to the largest slave power in history.

It's tempting today to see people dying or rain forests disappearing far away and say, "what I do doesn't matter," but we are driving the process. We are who matters. The English person in 1800 who bought slave-produced sugar and tea didn't whip or shackle anyone in chains, but they funded the plantations. Germans who in the 1930s didn't commit any violence by paying their taxes still funded the rise of Nazism.

As long as we procrastinate, we are the English in 1800 and the

Germans in 1935. I'm not saying we are like Nazis, in fact the opposite. We are like non-Nazis living in Germany. Just as few of us want to pollute, "most Germans did not want the Jews to be killed. Many ordinary Germans even provided Jews with understanding and support."[41] Still, few engaged. Most procrastinated: "It took nearly the entire German population to carry out the Holocaust." In fairness, PAID culture has refined Nazi propaganda to influence us more effectively today, though we call it advertising and public relations.

Still, we don't *feel* like we're supporting slavery and fascism. Why not?

41   Eric A. Johnson, *Nazi Terror: The Gestapo, Jews, and Ordinary Germans* (2000).

CHAPTER 13

THE PROBLEM IS US

It's one thing for Williams and Kendi to say slavery causes racism in history books or that complacency in implementing competitive strategies can be fatal and cause cruelty and violence, but what actually happens in a person's mind and heart? How does someone, once an innocent child, become cruel and violent, then say that they love the slaves they deprive of freedom and torture and that the slaves love them back?

Slaveholders acting against their values drove patterns identical to the death spirals of chapter 7 that drove their inner elephants to transform their riders beyond mere PR-firm mode to PR-firm-on-steroids mode. Their first death spiral was their internal one, which preceded our death spirals I described above by centuries. Slaveholders in the American South considered themselves good, but knew they weren't innocent, creating internal conflict. James Oakes' 🎤 *The Ruling Race*, Manisha Sinha's 🎤 *The Counterrevolution of Slavery*, and William Sumner Jenkins' *Pro-slavery Thought in the Old South* documented slaveholding culture from their perspective. Here are the words of a "Louisiana mistress" stating feelings she could not escape:

> Always I felt the moral guilt of it . . . how impossible it must be for an owner of slaves to win his way to heaven . . . My soul hath felt the awful weight of sin, so as to despair in agony—so as to desire that I had never had being. Oh God! . . . to suffer under the burden of my guilt.

They knew their actions violated what they believed was right. Oakes wrote, "A Virginia master echoed these sentiments. 'This, sir, is a Christian community,' he wrote in 1832. 'Southerners read in their Bibles, *Do unto all men as you would have them do unto you*; and this golden rule and slavery are hard to reconcile.'" Note the psychology in it being hard to *reconcile*, not hard to *stop*. They didn't want to end the external conflict. They wanted to reconcile their internal conflict.

The *Richmond Enquirer* wrote in 1856:

> In this country alone does perfect equality of civil and social privilege exist among the white population, and it exists solely because we have black slaves . . . Freedom is not possible without slavery . . . The abolition of negro slavery in the South would inevitably end in the ruin of the political constitution of the country.

It bears repeating. They said **"Freedom is not possible without slavery."** Vice President John Calhoun stated on the senate floor in 1837: "The relation now existing in the slaveholding States between the two [races], is, instead of an evil, a good—a positive good."

William Harper, representative from South Carolina, wrote,

> Slavery has done more to elevate a degraded race in the scale of humanity; to tame the savage; to civilize the barbarous; to soften the ferocious; to enlighten the ignorant, and to spread the blessings of Christianity among the heathen, than all the missionaries that philanthropy and religion have ever sent forth.

A northerner who went south saying he had "every prejudice I could have against slavery" tried to resist enslaving people. He claimed he couldn't help it: "Servants I want; it is lawful for me to have them; but hired ones I cannot obtain, and therefore I have purchased some . . . I cannot do as I would, I do as I can." To me, these words sound like words I hear today: "What can I do? What I do doesn't matter."

When they harmed slaves and deprived them of their freedom, slave-holders may have tried to convince themselves that Africans were less than human, that they were civilizing them, and so on. Seeing evidence of their humanity would expose the lie, create internal conflict, and prompt anger and hatred toward the slaves, leading them to punish them more, driving the internal death spiral.

INTERNAL CONFLICT DEATH SPIRAL

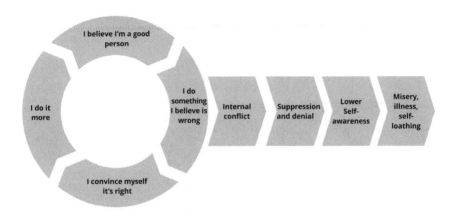

The second, external death spiral, led slaveholders to hate the people they hurt for being human when they wanted to believe they weren't. Seeing the slaves' suffering caused them to face their guilt, shame, and other results of corruption. They might think: "I didn't feel guilty. Then I interacted with you. Now I feel guilty. You caused it." This death spiral resulted in racism.

EXTERNAL CONFLICT DEATH SPIRAL

Pollutionism today plays the role racism did (and does). In the next chapter, we'll see I'm not just saying pollutionism is analogous to racism; they are the *same* parts of the *same* dominance hierarchy.

The third death spiral occurred too, this one cultural. When enough people felt the first two spirals, individual feelings became culture. Enough people said that slavery was right that it became a cultural norm that they stopped questioning.

Cultural acceptance served high-status people by decreasing internal conflict that could create civil unrest. Authorities supported "proven" results, creating what we now call scientific racism to institutionalize racist beliefs. To someone with a conscience wracked with guilt and shame, it seemed plausible.

CULTURAL CONFLICT DEATH SPIRAL

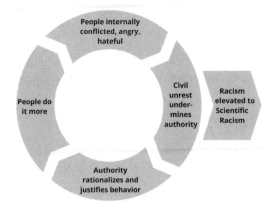

They did "scientific" things like measure cranial sizes to "prove" Europeans had bigger brains so were smarter than Africans. The measurements were accurate. The theory seemed plausible. Scientific racism eased their consciences and became an unquestioned part of culture. This process is likely how Jefferson contrived to believe Africans mated with orangutans and his other racist beliefs devoid of evidence. To us, his beliefs sound shameful, even stupid. Future generations will see our scientific pollutionist beliefs as shameful, even stupid too. People post-mindset shift already do. In fairness, as much as these beliefs look shameful, they emerge from the same processes I saw at the Chelsea Fridge. We're humans responding to our conditions.

We call scientific racism pseudoscience because in science we don't start with a conclusion and look for evidence to justify it. Scientific racism did so to mollify feelings of guilt and shame. The Nobel Prize-winning scientist Richard Feynman said about theories:

> Compare it directly with observation to see if it works. If it disagrees with experiment, it's *wrong*. In that simple statement is the key to science. It doesn't make a difference how beautiful your guess is, it doesn't make any difference how smart you are, who made the guess, or what his name is, if it disagrees with experiment, it's wrong. That's all there is to it.

Scientific racism is wrong in Feynman's sense: it's inconsistent with observation. Nature keeps conflicting with its predictions. Science works by starting with nature and looking for conclusions. Scientific racism started with the conclusion and looked to support it. Today we see scientific racism as shameful lies, but many felt they honestly believed it, or tried to, because they found it affirmed their lifestyle. It felt warm and supportive.

Scientific pollutionism today plays the role scientific racism did then. It promotes "solutions" that are the equivalent of past "solutions" to slavery that weren't designed to work but to soothe the consciences of slaveholders and even opponents of slavery who didn't act. They promoted colonization (shipping African-Americans to colonies in Africa or elsewhere), expanding slavery to new territories to diffuse it away, and gradual emancipation that kept getting more delayed. The energy "solutions" listed in the coda that don't work are today's equivalents of this thinking.

Future generations will dismiss scientific pollutionism as shameful bunk, but it appeals to us. It enables us to believe we are helping those we hurt. It tells us, "you're a good person."

As with scientific racism, scientific pollutionist "evidence" and "proofs" don't need to be right either. We control the resources so we can dominate. We only have to ease our consciences enough to sleep at night and keep using killing and raping as ways to gain power, comfort, and convenience. The more we accept our self-serving claims, be they pollutionist or racist, the more people in the community support each other and the less we question our lies and lives. The expectations become self-fulfilling. For example, we claim individual actions don't matter, then avoid straws for a week or some other minimal action, and "conclude": "See, individual actions don't matter."

The disdain and repugnance we look back on believers and proponents of scientific racism with will be how future generations will see our believing scientific pollutionism. Beth's recommendation letter shows that graduates of Spodek Method workshops already do.

Comparing Scientific Racism With Scientific Pollutionism

Scientific racists didn't say, "We're making stuff up to make you feel better about hurting people." They said, "Here's the science. It's not our opinion. It's fact." Scientific pollutionists don't say they're making stuff up either. They say "Here's the science. It's not our opinions. It's fact." Their process of "proving" racial inequality is our process of proving decoupling, green growth, and what created peace today will maintain peace tomorrow.

Scientific racism and scientific pollutionism are popular not because they are right but because they comfort us. They contort science to "prove" themselves right. One solution is to remain humble to nature. Another is to learn from history. We learn about Abraham Lincoln and Oskar Schindler because we aspire to be like them. But we whose planes and cars spew poisonous exhaust and who fill garbage bags with plastic every week aren't like them today. If we want to learn about ourselves, we should read Oakes's, Sinha's, and Jenkins's books on slaveholders' culture to see how other people with status in a dominance hierarchy and didn't oppose it felt and thought.

You can find examples of scientific pollutionism in the work of Steven Pinker, Julian Simon, Bill Gates, Elon Musk, and other ecomodernists and techno-optimists.

You don't have to take my word for it. Scientific pollutionists sharing their beliefs shows how much they resemble scientific racism. The same psychological processes drive them.

FOUNDATION FOR ECONOMIC EDUCATION (TODAY)	EDWARD BROWN, SOUTH CAROLINA LAWYER (1826)
Nothing has done more to lift humanity out of poverty than the market economy. This claim is true whether we are looking at a time span of decades or of centuries . . . If we look over the longer historical period, we can see that the trends today are just the continuation of capitalism's victories in beating back poverty. For most of human history, we lived in a world of a few haves and lots of have-nots. That slowly began to change with the advent of capitalism and the Industrial Revolution. As economic growth took off and spread throughout the population, it created our own world in the West in which there are a whole bunch of haves and a few have-more-and-betters.	Slavery has ever been the stepping ladder by which countries have passed from barbarism to civilization. History, both ancient and modern, fully confirms this position. It appears, indeed, to be the only state capable of bringing the love of independence and of ease, inherent to man, to the discipline and shelter necessary to his physical wants.

and:

ENLIGHTENMENT NOW, STEVEN PINKER	SLAVEHOLDERS
Ecomodernism begins with the realization that some degree of pollution is an inescapable consequence of the Second Law of Thermodynamics. When people use energy to create a zone of structure in their bodies and homes, they must increase entropy elsewhere in the environment in the form of waste, pollution, and other forms of disorder. The human species has always been ingenious at doing this—that's what differentiates us from other mammals—and it has never lived in harmony with the environment . . . A second realization of the ecomodernist movement is that industrialization has been good for humanity. It has fed billions, doubled life spans, slashed extreme poverty, and, by replacing muscle with machinery, made it easier to end slavery, emancipate women, and educate children. It has allowed people to read at night, live where they want, stay warm in winter, see the world, and multiply human contact.	"[Slavery] has existed and ever will exist, in all ages, in some form, and some degree. I think slavery as much a correlative of liberty as cold is of heat. History, experience, observation, and reason, have taught me, that the torch of liberty has ever burnt the brightest when surrounded by the dark and filthy, yet nutritious [influence] of slavery." — W. H. Roane, *Richmond Enquirer* (1831) "I do not hesitate for a moment in maintaining that the slave trade has been a source of incalculable blessing to mankind. Just so far as African Slavery in the United States is superior to African Slavery as it exists in Africa—just so much good has resulted from the slave trade." — Matthew Estes, *A Defense of Negro Slavery As It Exists in the United States* (1846)

and (to clarify, I'm comparing how Epstein characterizes the system, not criticizing the workers employed by the system who were born into it; they didn't create it):

ALEX EPSTEIN IN *THE MORAL CASE FOR FOSSIL FUELS* (2014)	CONGRESSMAN JAMES HENRY HAMMOND (1836)
There is a group of people who are working every day to make sure that the machines that can make us safe from our naturally dangerous climate and enable us to thrive in it have all the energy they need. These people work in coal mines, on oil rigs, in laboratories, in boardrooms, all devoted to figuring out how to produce plentiful, reliable energy at prices you can afford—because that is what their well-being depends on and, in my experience, because they believe that it is the right thing to do. Those are the people in the fossil fuel industry, who are dehumanized in the media on a daily basis, who are tarred as Big Oil or, in the case of workers, such as coal miners, are portrayed as dupes who don't know what they're doing, who aren't wise enough to know they're making our climate unlivable through the work that supports themselves and their families.	Slavery is said to be an evil . . . But is no evil. On the contrary, I believe it to be the greatest of all the great blessings which a kind Providence has bestowed upon our glorious region . . . As a class, I say it boldly; there is not a happier, more contented race upon the face of the earth . . . Lightly tasked, well clothed, well fed—far better than the free laborers of any country in the world . . . their lives and persons protected by the law, all their sufferings alleviated by the kindest and most interested care. Sir, I do firmly believe that domestic slavery regulated as ours is produces the highest toned, the purest, best organization of society that has ever existed on the face of the earth.

By contrast, compare views of nature between PAID culture and a Sustainable Free Abundant one:

ALEX EPSTEIN IN *THE MORAL CASE FOR FOSSIL FUELS* (2014)	PAULA GUNN ALLEN, LAGUNA PUEBLO (1979)
One crucial truth is that climate is naturally volatile and dangerous. Absent a modern, developed civilization, any climate will frequently overwhelm human beings with climate-related risks—extreme heat, extreme cold, storms, floods—or underwhelm human beings with climate-related benefits (insufficient rainfall, insufficient warmth). Primitive peoples prayed so fervently to climate gods because they were almost totally at the mercy of the naturally volatile, dangerous climate system.	We are the land . . . that is the fundamental idea embedded in Native American life . . . the earth is the mind of the people as we are the mind of the earth. The land is not really the place (separate from ourselves) where we act out the drama of our isolate destinies. It is not a means of survival, a setting for our affairs . . . It is rather a part of our being, dynamic, significant, real. It is our self . . . It is not a matter of being 'close to nature' . . . The earth is, in a very real sense, the same as our self (or selves).

Milton Friedman claimed we *depend* on pollution: "Even the most ardent environmentalist doesn't really want to stop pollution. If he thinks about it, and doesn't just talk about it, he wants to have the right amount of pollution. We can't really afford to eliminate it—not without abandoning all the benefits of technology that we not only enjoy but on which we depend." Yet when liberated from that dependence—addiction, really—in head-to-head competition, people consistently choose to remain free of it.

Regarding theories that economies need to grow or that we can't stop polluting, as Feynman would say, those theories are wrong. Countless human cultures have thrived without growing, polluting, or depleting. For example, some time after Polynesians discovered and made homes in Hawaii, trade stopped and those on the Hawaiian islands lived on their own for at least five hundred years, more than enough time to exhaust the land, plants, and animals if they lived unsustainably. Instead, they lived with a fixed amount of land for longer than the period from the Enlightenment to now. The Hawaiians were not unique in having lived sustainably that long. The Hadza lived tens of thousands of years sustainably and the San possibly hundreds of thousands of years. Human economies don't need to grow, pollute, or deplete. To believe otherwise is PR-firm-on-steroids talk.

CHAPTER 14

THE SYSTEM
OF SLAVERY
PERSISTS TODAY

Every nation has made slavery illegal. Despite loopholes and lax enforcement in some places, global culture has changed. Did this cultural
evolution from egalitarian to scientific racism end?

Or are racism and scientific racism connected to pollutionism and
scientific pollutionism? It turns out pollutionists themselves already
showed that they are, illustrated by a quote from University of Michigan
professor Andrew Hoffman. Writing on pollution and slavery: "The first
time these two concepts were linked for me was seven years ago, when a
senior oil industry executive in London asked me a rhetorical question:
'If it wasn't for oil, where would we get our energy?'"[42]

Andrew continued, "His answer, to my astonishment, was 'slavery.'"

The executive was saying that his industry helped end slavery. He
wasn't alone among oil executives with that belief. After all, they felt if
you could force a person to operate a cotton gin or to power it with a
steam engine, you'd pick the engine. As far as they knew, they replaced
suffering with benign technology. He filled in a missing step in the stages
of how we get energy:

---

42 Andrew Hoffman, "Climate Change: Calling the Fossil Fuel Abolitionists," *Ethical Corporation*,
    May 28, 2008.

edible plants and wood → draft animals →
coal, oil, gas → fission (→ fusion?)

to include a source of industrial energy:

edible plants and wood → draft animals → **slaves** →
coal, oil, gas → fission (→ fusion?)

He wasn't saying that pollution and depletion were *like* slavery. The connection was in the underlying dominance hierarchy. *They are one system.*

The system of slavery became today's system of pollution and depletion. We made slavery illegal, but the *system* only looked like it went away. The suffering appeared to decrease for two hundred years because it took that long to fill earth's reservoirs. Until then, pollution and depletion levels were too low to hurt many people directly. Even then people could reasonably believe the solution to pollution was dilution.

We now see that only the *mechanism* of cruelty changed. PAID culture is killing more than slavery did. Sadly, we are more than mere members of the home market driving this system. Many of us are today's equivalent of plantation owners and workers. Today's airports, power plants, factories, and container ships are our plantations. Today's flights, cars, doof, and bottled water are our sugar and cotton. We may not employ slaves ourselves, but we fund greater suffering,

Slavery Persists Today. PAID Culture Funds It.

We also fund slavery outright. Several podcast guests described slavery they witnessed. Ron Schoonover 🎙 worked for the US National Intelligence Council and the State Department for over ten years, becoming an expert on global ecological disruption and its impacts on people and societies. On the podcast, he described horrors resulting from environmental problems:

In 2014, '15, '16 and '19 and some years in between there was a shift in the manner of migration . . . It was no longer people out of work. It was more and more families and unaccompanied children.

They're being tortured, they're starving to death, they're being shot, they're homeless. We're talking about starvation. Talking about violence toward women and children. We're talking about rape. We're talking about targeted violence toward migrants.

These families, these children, were escaping, largely amplified by environmental change, the collapse of agriculture, and how it connected with the environment.

Siddharth Kara 🎤 wrote in his exposé *Cobalt Red*,

> Across twenty-one years of research into slavery and child labor, I have never seen more extreme predation for profit than I witnessed at the global cobalt supply chains. The titanic companies that sell products containing Congolese cobalt are worth trillions, yet the people who dig their cobalt out of the ground eke out a base existence characterized by extreme poverty and immense suffering. They exist at the edge of human life in an environment that is treated like a toxic dumping ground.

He documents teens killed for negotiating, girls and women raped, and miners dying in collapses. He continues, "Apple, Samsung, Google, Microsoft, Dell, LTC, Huawei, Tesla, Ford, General Motors, BMW, and Daimler-Chrysler are just some of the companies that buy some, most or all of their cobalt from the DRC." When we buy from them, whether we want to or not, we fund slavery.

Katie Redford 🎤 is a lawyer whose case *Doe v. Unocal* was "The First Case of its Kind: Holding a US company responsible for rape, murder, and forced labor in Myanmar."[43] Similar cases were brought against Shell, Chevron, Nestle, Coca-Cola, Rio Tinto, and peers.

---

43 "Doe v. Unocal," EarthRights International.

In other words, the mechanism for cruelty is different in PAID culture compared to the system of slavery, but cruelty is just as essential to PAID culture, which affects more people and is growing.

We may say we can't help participating in our world. So did Jefferson and Madison. We may insist we have no choice. So did they. Imperialism and colonialism today don't look like Roman legions and slave ships. They look like SUVs, car-dependent suburbs, and flying-distance families.

I point this connection out not to accuse or make anyone feel bad, but because a small number of people organized abolitionism and anti-slavery politics to make slavery illegal and succeeded. We can continue their legacy, though it requires us to see that within PAID culture, polluting and depleting technology didn't stop cruelty, it accelerates cruelty. It will continue to accelerate it until we change our culture.

CHAPTER 15

WE CAN CHANGE CULTURE. PEOPLE HAVE BEFORE.

Tyranny and other dominance hierarchies don't last forever. Opposing them takes work, but history contains examples restoring freedom: the development of democracy, jubilee years forgiving debt and freeing slaves, the Magna Carta, abandoning divinity of rulers, creating antitrust law, extending the vote, creating public education, and writing our Constitution.

We can learn from what worked. Let's consider examples most relevant to sustainability.

Abolitionism

Abolitionism was a rare movement where one group worked for a *different* group's freedom instead of benefiting by depriving them of it. The two most important facts about abolitionism are, first: it succeeded. It changed a part of global culture over ten thousand years old to every nation passing laws against it. Second: it failed. Tens of millions of people remain in forced labor, sexual slavery, the US prison-industrial complex, and other forms of slavery. As we saw in part 3, the system still exists in PAID culture.

What Can We Learn From Abolitionism?

Books on abolitionism can fill libraries, so I can't cover everything, but I'll share a few relevant points we can learn from it.

Though some consider the ending of slavery inevitable, it seemed inconceivable for most of history. Adam Hochschild 🎤 wrote in *Bury the Chains: Prophets and Rebels in the Fight to Free an Empire's Slaves*, "At the end of the eighteenth century, well over three quarters of all people alive were in bondage of one kind or another, not the captivity of striped prison uniforms, but of various systems of slavery or serfdom."[44] Many benefiting from it saw it as benign as we see activities enabled by polluting and depleting today: "The era was one when . . . 'freedom, not slavery, was the peculiar institution.' This world of bondage seemed all the more normal then, because anyone looking back in time would have seen little but other slave systems" everywhere globally.

Slaves rebelled since before Spartacus, but a movement to end the institution only began in Philadelphia (near where I grew up), before the American Revolution, then spread to Europe. As abolitionism grew in England, many predicted that if they banned slavery, other slave-trading empires would pick up the lost trade. They said their empire would collapse with no decrease in slavery. The opposite happened. After England banned the slave trade, France, Spain, Portugal, Holland, and others followed. None collapsed.

Industries benefiting from slavery predicted losing business, such as English iron workers, who manufactured shackles and weapons for slave-traders. Instead, abolition improved their businesses, as trade increased without slavery.

Abolitionism provides role models, including Toussaint L'Ouverture, Harriet Tubman, Frederick Douglass, and Sojourner Truth, but they came from the oppressed group. I hope leaders emerge from today's oppressed groups, but most of us are members of PAID culture and *support* oppressive

---

44 Adam Hochschild, *Bury the Chains* (2005).

industries, even if we have low status. We need to find role models who *could* have participated in oppression but didn't. That is, since our equivalents are the English sugar consumers in 1800 and German tax-paying citizens of 1935, whom among them can we model ourselves on?

One candidate is Robert Carter III, who freed five hundred slaves, but he didn't lead others. Benjamin Franklin freed slaves and became president of the Pennsylvania Abolition Society, but held back from influencing the Constitution. Washington, Jefferson, and Madison talked big on freedom, but on slavery they capitulated (for that matter, Douglass demeaned Catholics and Native Americans; when he held political office, he promoted American imperialism). John Brown and William Lloyd Garrison dedicated their lives to abolitionism so I recommend learning more about them, but I'll focus on three.

The first two are leaders of British abolitionism: Thomas Clarkson and William Wilberforce. Their transformations and resolve inspired generations of reformers, including me. Their stories, as Adam Hochschild told in *Bury the Chains*, revealed to me the potential for global change in decades. Eric Metaxas's 🎤 book *Amazing Grace* focused on Wilberforce. I suspect the Trump-supporting Metaxas and *Mother Jones* co-founder Hochschild might not expect to be mentioned together, but I spoke with each on my podcast. Their agreement helped me see that however polarizing abolition was, it unites us today. I foresee sustainability uniting us and future generations.

Clarkson learned of the cruelties of slavery inadvertently, researching for a student essay contest at Cambridge University. Heading home after winning, slavery "wholly engrossed my thoughts" leading to his mindset shift:

> I stopped my horse occasionally, and dismounted and walked. I frequently tried to persuade myself in these intervals that the contents of my Essay could not be true. The more however I reflected upon them . . . the more I gave them credit . . . I sat down disconsolate

on the turf by the roadside and held my horse. Here a thought came into my mind, that if the contents of the Essay were true, it was time some person should see these calamities to their end.

Wilberforce experienced his mindset shift, writing:

So enormous, so dreadful, so irremediable did the [slave] trade's wickedness appear that my own mind was completely made up for abolition. Let the consequences be what they would: I from this time determined that I would never rest until I had effected its abolition.

The third role model is a leader of American abolitionism, Abraham Lincoln. Books on him fill libraries so I won't try to cover him comprehensively. I'll share some points I found helpful. In short, he combined the words of Jefferson with the action of Tubman and integrity of Carter.

He was willing to learn and evolve. He focused on working with teams. Before about 1862 he would have bristled at being called an abolitionist. After 1863 he would have admitted his debt to the "old movement," crediting a long history of stages from abolitionists as early as the 1730s, more known with Garrison and the American Anti-Slavery Society, ramping up a century later. He knew the movement transformed from a small group of radical outsiders into a political movement including his party.

He realized that a house divided against itself could not stand—that is, a Constitution that allowed slavery anywhere was too internally conflicted to protect freedom. He tried to influence slaveholders and their state governments, but they wouldn't yield. He realized that legislation, judicial decisions, executive orders, technology, or market reforms wouldn't end slavery. Even if the Emancipation Proclamation had freed all slaves, the Civil War could have ended with the institution of slavery intact and slave states could have rebuilt their empire. He saw that ending slavery meant not just freeing existing slaves. Slaves weren't the problem. Slaveholders were. He had to amend the Constitution.

It's worth reviewing how the Thirteenth Amendment came to pass. The first mention of a constitutional amendment to end slavery came in 1827, in *Freedom's Journal*, the first African-American newspaper. The idea must have seemed ludicrous. It would need three-quarters of the states to ratify it, but half the states were pro-slavery, and they included the richest regions and people in the world. Even nearly forty years later, after the Civil War, no one knew if it could pass.

Many abolitionists may not have lived long enough to see their work make the amendment possible, yet such a close vote meant that it may not have passed without each of their efforts. Every contribution mattered. Still, we can speculate: had more people acted then, the amendment might have passed sooner. What if more of us act today?

The amendment helped turn the most divisive issue in American history into one of unity. Lincoln didn't do it alone, but he did everything he could to pass it. We don't all have to go to the lengths he did, but any of us can. There was only one Lincoln, after all, and plenty of abolitionists and anti-slavery politicians. Imagine if there were more Lincolns today.

I recommend learning about Lincoln and abolitionism. I've brought several renowned scholars of the fields to my podcast, including two two-time Lincoln Prize honorees, James Oakes and David Blight 🎙, as well as Manisha Sinha and John Brooke 🎙. I recommend listening to their episodes and reading their books, as well as those of Eric Foner, Sean Wilentz, and Doris Kearns Goodwin.

At Least Try

I believe sustainability needs Clarksons, Wilberforces, and Lincolns. Until someone does it better, I'm doing it the best I can and hope you'll join me.

Doing the best I can reminds me of a story from playing ultimate Frisbee in college. A teammate threw me a pass in a game. It was out of my reach so I watched the disc go past without making a play for it.

When I came off the field after the point ended, another teammate asked, "Why didn't you try for that pass?"

I said, "It was out of my reach."

He said, "At least try."

Since then I've always felt: "at least try." It means something to me when, say, in baseball someone hits a home run but the outfielder still runs to the wall and jumps for it. He knows he can't catch it, but he shows he's doing all he can. He leads the rest of his team. Everyone plays better, *and sometimes he catches it*. Basketball great Larry Bird said, "It makes me sick when I see a guy just stare at a loose ball and watch it go out of bounds."

"At least try" motivated me years later to run sprints alone in the rain to help the team make nationals. Do I like running in the rain? Well, we made nationals but lost in the quarterfinals. I wish I'd run more sprints.

Am I going to watch the system of slavery that never died cause suffering and threaten billions of lives when successful strategies, tactics, and role models lie unused? Will I become today's Abraham Lincoln? If it serves the mission, I'll at least try.

What Can We Learn From Abolitionists and Anti-Slavery Politicians?

Why didn't more free Americans oppose slavery? Understanding why helps us see why more accept PAID culture today.

Most free Americans benefited from slavery. Few actively opposed it. The rest may not have been virulent advocates, but most free northerners were racist—not because blacks were inferior, but because they weren't. Their PR-firm-on-steroids riders told them they were. Likewise, we today aren't pollutionist from evidence supporting pollutionist beliefs. We are pollutionist because those beliefs are false, and therefore our PR-firm-on-steroids riders try to convince us of them.

More of us can be like abolitionists and anti-slavery politicians. Those of us who devote ourselves at Lincoln-levels will love the experience. If you don't want to devote yourself that much, I hope you'll at least support those of us who do.

Ending slavery looked insurmountable. The arguments and

motivations to look the other way must have felt overwhelming. We might feel overwhelmed by resistance today, but we have advantages abolitionists didn't: we can learn from them.

What Can We Learn From Slaveholders?

Many saw the Civil War coming. If they had known its magnitude, even the richest and those who felt most secure may have realized even they would suffer.

If I were magically transported back to 1855 into the body of a slaveholder in the deep south, I think I can safely say I would not practice slavery and rationalize and justify it with racist or pollutionist excuses. I would free them as soon as practical. I believe you would too, even if you knew you couldn't change anyone else, even if your neighbors resented you for it.

Even if you were transported to 1855 but not a slaveholder, you would see the problems with buying slave-produced goods. Am I correct to believe you wouldn't practice slavery and rationalize and justify it with racist or pollutionist excuses either?

Other Role-Model Movements

Abolitionism is one example of change that happened within decades on the scale of empires, but it was long ago. Technology was different. What about today? Can America and other rich nations change?

Post-War Japan

In 1945, Japan had been destroyed. Many felt humiliated by unconditional surrender. Could the nation recover? Many leaders and managers were dead. The nation was known for producing cheap junk.

In 1950 several industry executives invited an American, W. Edwards Deming, to help. He had little authority or charisma. He wasn't handsome. He shared techniques he had developed and refined before and during the war. Despite Japan's being insular, they learned from him and acted on what he shared.

His results?

"Deming told the Japanese managers that if they followed his recommendations 'they would capture markets the world over within five years.' In fact, it began happening within four years!" wrote an engineer who profiled him.[45] The Union of Japanese Scientists and Engineers named a prize after Deming. Its recipient list "reads like a *Who's Who* of Japanese companies that have dealt serious blows to US industry . . . Winners include Toyota . . . Komatsu . . . Ricoh . . . Toshiba . . . Bridgestone, and Matsushita." In 1960, the Emperor awarded him Japan's Order of the Sacred Treasure, among Japan's highest honors.

No one person transforms an empire, but Deming played a major role in transforming an empire to supplying the world with high-quality affordable products in a few years. He documented his systems-based approach in his books, *Out of the Crisis* and *The New Economics*.

His humble background illustrates that many of us can achieve as much. His first job after university was with the US census in the 1930s. Before him, they tried to poll every individual. He implemented statistics in a specific way to sample the population that reduced error with less work. Engineers say, "Quality, time, price. Pick two," but his methods improved all three, and employee morale too.

During the war, he applied these techniques to building tanks and planes. Before him, they had to choose between quality and time. Deming applied statistical sampling to the manufacturing process, achieving again lower costs, higher quality, faster output, and higher morale. No one had dreamed such results. Like other teams developing radar and cracking the Enigma machine, his results off the battlefield helped win the war.

The men fighting abroad who had run the factories didn't know his techniques. When they returned, they squeezed him out and restored

---

45 Robert Austenfeld Jr., "W. Edwards Deming: The Story of a Truly Remarkable Person," *Papers of the Research Society of Commerce and Economics* 41, no. 1, (May 10, 2001).

their techniques. He learned that cultural change required support from industry and government leadership. Hence, in Japan, he insisted on including CEOs, owners, and high-level government officials.

After Japan, Deming returned to the US, where American industry still didn't recognize him. Then a documentary in 1980 called *If Japan Can, Why Can't We?* featured his results there. Suddenly American industries demanded him. Restarting his career in his 80s, he helped transform these industries. President Reagan awarded him the National Medal of Technology in 1987. The National Academy of Sciences gave him the Distinguished Career in Science award in 1988.

When asked, toward the end of his life, how he wished to be remembered in the US, he replied, "Maybe as someone who spent his life trying to keep America from committing suicide."

Corporal Punishment of Children

My parents spanked me growing up, though only a few times. It seemed normal then, though less than in my parents' childhoods. My mom's father locked the children in the basement. Her mom was strict and hit the children too. I thought the decreasing trend in corporal punishment of children just happened. As it turns out, that trend resulted from deliberate action.

Historically: "For millennia, corporal punishment of children has been justified by religious texts, criminal laws, and cultural beliefs in the power of pain to engender learning. It remains one of few forms of violence that is still accepted, normalized, and prescribed, albeit to varying degrees across cultures."[46] Scripture supports it, such as, "He that spareth the rod, hateth his son; but he that loveth him, chasteneth him betimes," and "Foolishness is bound in the heart of a child; but the rod of correction shall drive it from him."

---

46 Joan E. Durrant, "Corporal Punishment: From Ancient History to Global Progress," *Handbook of Interpersonal Violence and Abuse across the Lifespan* (October 13, 2021): 343–365.

No clear point changed the trajectory, but in the seventeenth century, John Locke "introduced new ideas about childhood, including the notion of 'tabula rasa' which diverged sharply from the concept of original sin. He considered beating of children a form of tyranny and urged parents to replace it with reasoning and shaming." Then "in eighteenth-century France and England, some prominent thinkers argued against the brutality of 'scholastic punishment' and questioned the idea that children deserved institutionalized humiliation," and "In 1839, the first parenting book to oppose all corporal punishment of children was published."

By today, "as a result of the twentieth century's recognition of children as persons and rights-holders, the world has undergone profound change in recent decades, with respect to corporal punishment . . . To date, 59 countries have fully prohibited corporal punishment in all settings, including the home."

Notably, "there is considerable evidence that corporal punishment of children is rarely found among those hunter-gatherer societies classified as egalitarian, which tend to share social norms that do not condone violence." This pattern is consistent with corporal punishment reinforcing hierarchy: "Status continued to serve as a determining factor in the application of corporal punishment."[47]

Reducing corporal punishment of children may be less a recent or Enlightenment trend so much as restoring a Sustainable Free Abundant cultural value. I see it as a role model movement.

Corporate Role Models

At the company level, Ray Anderson, the founder and CEO of Interface Carpet, faced many objections when he decided to make the company more sustainable. He overcame them, leading to his company thriving while implementing many sustainability practices. The corporate culture

---

47 Guy Geltner, "History of Corporal Punishment," *Encyclopedia of Criminology and Criminal Justice* (November 27, 2018): 2016–2115).

transformed to embracing sustainability. When he died, the next CEO first intended to return Interface to mainstream business practices, but when he learned of the results, he kept it going.

When CVS drug stores rebranded to CVS Health, they realized selling cigarettes conflicted with its new identity. They considered if they should stop selling them, but cigarettes were their most profitable product. Nearly all financial and economic advice suggested to keep selling them. Instead, they chose to stop. Within a year, their profits surpassed previous levels and their competitors are considering following. The head of a branding agency involved with the project, Mark DiMassimo 🎙 shared on my podcast the personal challenges executives shared in this decision.

The B-corp movement faces similar objections. Companies that become B-corps choose to write their by-laws to include employees, communities, and the environment in decisions that affect them. These considerations restrict them from options their competitors retain. Common sense would suggest that a company with more options would beat B-corps, yet the B-corp movement is growing and thriving. Lorna Davis led the largest transition to a B-corp so far with Danone North America, a wholly owned subsidiary of Danone, which is a global publicly traded company with annual revenues over forty million dollars.

City Role Models

Did you think Amsterdam was bicycle-friendly naturally, as a once-medieval city built for walking? I did. On the contrary, in the mid-twentieth century it was overrun with cars. Developers planned highways into the city center to bring in more cars in the style of Houston, once a walkable streetcar town. Citizens opposed the car-centered design. They overcame objections that if Amsterdam didn't accommodate more cars, it would lose stature or fail its businesses. They halted the highways and began Amsterdam's path to the safe, livable city we know today. Other Dutch cities followed, as have many cities worldwide.

Seoul, for example, began by demolishing an overpass in 2014, replacing it with bus lanes.[48] The process led to taking down more and more highways, revealing a creek it had covered, and adding a park around it.[49] The project improved traffic around the city, enjoying support from residents and tourists. It also helped the mayor who promoted it to become president of the nation.

When New York acted to ban smoking in the workplace in 2000, it had to overcome objections. People didn't know bars could have clean air. When they experienced it, they preferred it so much that New Jersey followed. I'm not aware of anyone proposing repealing the ban or even of any need for government to enforce the ban.

When New York City considered making large sections of Times Square pedestrian-only, local businesses opposed, saying they needed car traffic. The city overcame the objections. The results: more foot traffic, far exceeding car traffic in safety and business for local businesses. I don't know of anyone proposing reverting the pedestrian-only spaces to allow cars.

São Paulo, Brazil overcame similar objections in considering banning many forms of billboards. They enacted the ban. Five years later a survey "found that 70 percent of residents say the Clean City Law has been 'beneficial.'"[50] Cities around the world are following. I hope New York City does.

San Francisco residents knew its Embarcadero highway cut the city off from its waterfront. Studies showed that it exacerbated traffic congestion and recommended removing it. In the 1980s, the board of supervisors voted to tear it down. The mayor said, "I'm determined that that freeway come down, even if I have to become totally gray-haired in the process." Still, many feared removing it would increase congestion and their objections kept it in place. Then an earthquake damaged it in 1989, making it easier

---

48 Philippe Mesmer, "Seoul Demolishes Its Urban Expressways as City Planners Opt for Greener Schemes," *The Guardian*, March 13, 2014.

49 "How South Korea Demolished a Freeway to Restore Ancient River System into Green Urban Oasis Seoul," *The Primest*, November 7, 2022.

50 Cord Jefferson, "A Happy, Flourishing City With No Advertising," *Good.is*, December 26, 2011.

to remove, which they did. The result: renewed waterfront access, lowered congestion, and raised property values. Nobody wants it back.

Individual Role Models

Growing up, my main role models for leadership were Gandhi, Nelson Mandela, Martin Luther King, and Malcolm X. These role models still inspire me, but they aren't right for sustainability in PAID culture. They all came from oppressed groups and worked for their freedom. As much as we today fear suffering, *we* are polluting and depleting.

To help ourselves change our own culture, we need role models who could have benefited from oppressing others but didn't, then acted contrary to their culture. I mentioned Carter, Franklin, Clarkson, and Wilberforce. Sinha's *The Slave's Cause* recounts hundreds of abolitionists and antislavery politicians who could have owned slaves. We can view them as role models. They worked not for personal gain, to make quarterly numbers, or to increase sales, but to do what they thought was right.

In World War II, people like Oskar Schindler and Dietrich Bonhoeffer could have passed through the war with little personal risk. Neither acted to look good for others. They did what they thought was right. So did the family that hosted Anne Frank, among countless others who acted similarly. We today could learn from them.

Raoul Wallenberg was a Swedish businessman who went to Budapest in 1944 to help rescue Hungarian Jews from the Nazis and succeeded in saving thousands. Aristides de Sousa Mendes was a Portuguese diplomat in France who saved thousands of refugees fleeing the Nazis, including Salvador Dali. To do so, he disobeyed orders from the Portuguese dictator Salazar. Both Wallenberg and Sousa Mendes were born privileged and took personal risks to do what they believed was right. You can imagine the gut-checks they must have gone through to oppose Nazis, including Wallenberg arguing face-to-face with Adolf Eichman, but he helped save up to seventy thousand Jews. We can learn from them.

A recent role model and inspiration to me is Muhammad Ali for resisting being drafted to fight in Vietnam. No celebrity had yet spoken against the war when the army drafted him. He would gain materially if he accepted being drafted. The army would not risk sending him to battle, only to rally troops and promote the war. Even Jackie Robinson advised him to enlist. Instead, Ali resisted. In retaliation, the government revoked his passport and boxing license and prosecuted him. He risked his reputation, ability to fight at his physical prime, and bankruptcy to do what he felt was right.

Years later, the Supreme Court unanimously overturned his conviction. He opened the door for other influential people to speak publicly against the war, including Martin Luther King, who had held back from speaking publicly based on his relationship with President Johnson. Jackie Robinson recanted his advice.

Many people who overcame addiction and used their experience to help others have become role models, including podcast guests who overcame their addictions to become bestselling authors on the topic, Holly Whitaker 🎤 and Carl Erik Fisher.

PART 4

THE
SOLUTION

CHAPTER 16

ACT WITH RESOLVE, COURAGE, AND VISION

You may be wondering, "What if everyone stopped participating in our markets and buying things? Wouldn't society collapse?" As Deming found, we can keep the best of both worlds: restoring values we've lost and applying them to modernity, continually improving. I've written in this book about the risk of future collapse, but I'm not trying to solve a future problem. Some say we have to feel the pain before we act. Some say addicts have to hit rock bottom before they acknowledge their addiction and act.

I am solving problems here, now, affecting us directly, not later, elsewhere. The environmental problems that the media cover that are distant or projected for the future are abstract relative to our internal conflict resulting from our death spirals. If we aren't acting to our potentials, we can't look children in the eye and tell them we are doing our best to make a safer, healthier, more secure world for them. If we tell ourselves we are powerless, give up, and conform to PAID culture, we're actively making their worlds less safe, healthy, and secure. What is more important to us?

If we are not living sustainably but value the Golden Rule, we are violating our values, no matter how much our PR-firm-on-steroids rider tries to convince ourselves we're doing our best and to hide our inner torment. We are depriving people of life, liberty, property, and freedom. We are damaging ourselves. Solving our problems here and now is the best way to solve problems remote or projected in the future.

Starting a movement takes work and vision. It was hard for people boycotting buses in Montgomery, Alabama to imagine themselves succeeding, to walk when others said, "The bus is going anyway, just take it," or "the boycott won't work, you're walking for no reason." They may have felt like giving up, but I bet none regretted the effort. One seventy-two-year-old woman explained why she persisted: "My feets is tired but my soul is at rest."

Who could have expected Clarkson pondering on the side of the road his essay on slavery would, with a small community, lead an empire, then the world? He thought, "It was time some person should see these calamities to their end." Wilberforce joined, having resolved, "I would never rest until I had effected its abolition."

PAID culture will resist changing to one of everyone joyfully living sustainably. We will face headwinds, but we can resolve ourselves. We can mobilize a nation.

What was your biggest takeaway from Beth's recommendation on page 20? I'll tell you mine, but first some backstory: She wasn't looking to act in the field of sustainability. Her daughter Evelyn led her to it when she did the workshop. Evelyn wasn't looking for it either. She had contacted me for straight leadership coaching. When I mentioned sustainability, she felt satisfied for being part of a group she later described as "saying all the right things and buying all the right products for the environment." She took the workshop, but pushed back in sessions, saying she had the least knowledge and experience in the cohort.

When she tried the Spodek Method with her mom and friends, they resisted. She insisted, "If we're getting them to do something they weren't going to, isn't that coercion?"

Then the Spodek Method clicked. She saw it evoke emotions the person had been keeping inside and liberated them to act on these feelings—the opposite of coercion. She didn't need to know more science. Soon she began sharing how Spodek Method conversations improved her relationships with her sons and mom, like enabling them to talk about an issue they cared about without antagonism.

Beth is a retired grandmother. She worked seven decades so she could enjoy retirement, which for her meant flying around the world. She has family all over. She wants to share her love for them. The workshop led her to pursue figuring out how to achieve her goals while decreasing harming others. She's bringing her life in line with her values by actually acting.

Her experience is also why she's recommending something that made her cry and unearthed suppressed feelings of "gloom and doom," self-talk of helplessness, deprivation, and pessimism. Those feelings were like her "before" picture. She is now living "after," feeling joy, empowerment, connection, self-awareness, enthusiasm, and more.

Which leads to my biggest takeaway: **Beth transformed in seven weeks what took me seven years**, despite her starting without interest or intent, by focusing not on the y-intercept but the slope—that is, not on compliance, but enthusiasm. Now that she wants to act for her own reasons, she will overtake those who only comply for extrinsic reasons. Beth knows she will help her community and they'll thank her for it.

She is already leading others. Her sister Trish 🎤 took the workshop *while fighting terminal cancer*. If anyone could say, "I'll die before the worst happens, I deserve to indulge in my last moments," she could. It took me a while to understand why a cancer patient would work on sustainability leadership. First, someone aware of her mortality chooses deliberately. She is choosing what is most important to her. Second, she sees that our environmental problems aren't abstract or far away in time or place affecting others. They are here and now. The death spirals are hurting us in the moment, even those of us with fewer moments left. She is helping herself and those she loves, which is about as much as any of us can ever do. The Spodek Method enables it in an area most of us have given up on. She *is* indulging—in compassion, purpose, and love.

CHAPTER 17
LIFE SPIRALS

Humanity is like an orchestra preparing to play Carnegie Hall in a command performance. Everyone is looking forward with anticipation. The media is covering it.

One problem: no one in our orchestra has practiced their instrument. Everyone studied musical theory but never played a scale. Moreover, each insists that the performance is too big and too soon for individual practice. "Individual action doesn't matter," they say. "Each of us practicing solo doesn't scale to collective action." Everyone insists that the less time remains, the less time should be wasted on individual action.

Panic ensues. The players start breaking instruments. It's all they can do, not knowing how to play them, saying at least doing so makes some noise. They insist that geniuses will make instruments that will solve the problem. Anything but practicing.

This situation is our world. Nearly everyone, including top environmentalists, says we need to play as an orchestra but *not to practice our instruments*.

The way out of a death spiral is a life spiral, or virtuous cycle. How we reach sustainability will be how musicians reach Carnegie Hall: *practicing the basics*. Practicing the basics creates continual improvement—life spirals.

When I played ultimate Frisbee, teams I played on reached nationals and worlds. If you haven't seen the sport, each player has to throw the disc, which takes practice to do well. When I started, I feared causing turnovers with my then-poor throws, so I feared catching the disc since

I'd then have to throw it. I subconsciously tried *not* to get open, though, if you asked me, I would have said I was playing as best I could. I sabotaged myself and my team.

I practiced until my throws became good, then great. Then I wanted to play in the biggest points of the biggest games. I still knew I could make mistakes, but practice gave me confidence.

Sports and performing music are active, social, emotional, expressive, performance-based fields, as is leadership, including sustainability leadership. Mastering such fields goes through stages. First piano lessons often include instruction like, "Put your thumb on this key, your index finger on this key," and so on to teach the basic mechanics of the C major scale. The student practices basic motions. They may seem mechanical, but when the student is comfortable with them, the student can advance to other scales and greater comfort. Do we want to play mechanically forever? No, but we learn performance-based fields that way.

Practice enables us to advance from scales to simple musical pieces to complex pieces. Along the way we fail and learn from success and failure. With more practice we learn to express ourselves through the music. Eventually we may choose to learn to play for audiences. We connect with them and other artists. We increase our self-awareness, self-expression, and connection.

The more we practice, the more we improve, the more we develop as a person, and the more we enjoy practicing, creating a virtuous cycle, or a life spiral. Still, however far we go, we always benefit from practicing basics. Before winning Wimbledon, Serena Williams practiced the same ground strokes that I learned in my first lesson. LeBron James practices free throws before NBA finals. Yo-Yo Ma plays scales. Mick Jagger practices scales for hours before performances.

Sustainability leadership has something beyond any other active, social, emotional, expressive, performance-based field: everyone loves nature. I may have loved ultimate Frisbee to where I ran those sprints alone in the rain, but years before I cried when I was forced to practice violin.

People may like some fields and not others, but *everyone* has connected with nature in ways that created deep, powerfully motivating emotions, though some may hide it at first.

The Spodek Method is the best exercise for sustainability leadership I've found—actually the only one I know of. It's the C major scale, the ground stroke. Each of us has the opportunity to embrace sustainability and sustainability leadership. There is no limit to how far we can develop ourselves and others by mastering it, increasing our self-awareness, self ex pression, and connection in the process. I don't want to sound melodramatic, but the result is glory, not just personal glory, but glory in the art and performance, glory for humanity. Beyond any other field, virtuosity in sustainability leadership can also mean helping to reduce suffering and save lives, potentially of billions of people and countless species.

Gut Checks

I'm about to present a long-term solution many of you will react to as impossible or extreme, at least at first glance. Your mind will fill with objections. Mine did when I first conceived of it. Then, over time, it will make more and more sense as you see it resonates with your deepest values and solves our current problems.

What Are Your Values?

What are your deepest values? I ask because I'm going to propose a long-term solution to our environmental problems that likely meets your values better than any alternative, given our environmental situation. You may agree intellectually that my proposal meets your values, but you've spent decades with PAID culture pulling your elephant's rings to act against your values; decades of your PR-firm-on-steroids rider convincing you that acting against your values was good, right, and natural. You may be so used to giving up that you resist engaging in something that would improve your life, family, community, nation, and world.

Here are some examples of values that may reveal your values and help clarify them. Do you value the following? If so, my solution in the next chapter supports them.

- A safe, healthy future for your children?
- A safe, healthy future for yourself and your generation?
- Safety and security in your neighborhood?
- National security, honor, duty, and service?
- Limited government that protects life, liberty, and property?
- Free trade with a government to ensure a level playing field so a free market can allocate resources to the most effective innovators?
- Fairness and equality?
- To oppose racism, sexism, homophobia, and comparable discrimination?
- Human rights, civil rights, justice, and social justice?
- To hold corporations and government responsible?
- Science, engineering, and clean, green, renewable energy?
- To help the poor and helpless? Are you infuriated that the people who pollute most and benefit from it use their resources to protect themselves from the harm they cause?
- To see the results of your hard work?
- Human ingenuity and innovation?
- To expand beyond this one fragile planet?
- Leisure?
- To follow your religion and allow others to follow theirs?

CHAPTER 18

OUR WWII-LEVEL MOBILIZATION: A CONSTITUTIONAL AMENDMENT

I remind you in this chapter that I focus on an American view here not to exclude others, but because experience tells me that specifics convey more meaning than generalities. I know my culture best, so I can speak more specifically about it. Sadly, I see the US as a nation deep in PAID culture, maybe the deepest, therefore the most important nation to change. I expect that solutions that can work here will work more effectively and faster elsewhere. Several nations have already taken steps on the path I am about to present.

Past generations rallied nationally and globally to achieve what seemed impossible. We mobilized to fight Hitler, walk on the moon, end polio and smallpox, and abolish slavery. These challenges were hard but brought us together and spurred innovation. Innovation wasn't just technical, it was also social.

Looking back, it's hard to imagine how impossible these goals looked, certainly harder than anything I've done by lowering my environmental impact. Goals look easier after they've been achieved.

Our WWII-level mobilization will be a constitutional amendment in the style of the Thirteenth Amendment. I call it a "WWII-level mobilization" and not a "moonshot"—that is, a mobilization like President Kennedy launched to have a person walk on the moon within a decade—because of the stakes if we don't engage.

We can live sustainably and a specific goal is easier to achieve than a vague one. An amendment requires three-quarters of the states to ratify—that is, overwhelming public support. Past mobilizations and moonshots needed overwhelming support too. A clear goal helps people innovate and respond to naysayers with, "Just watch us." A movement to pass an amendment will activate communities like Silicon Valley, Washington DC, and academia to work at the stage of stewardship (recall from page 30) of changing culture. They are currently innovating *within* culture, which doesn't change culture and amounts to stepping on the gas, thinking it's the brake, wanting congratulations.

We've passed dozens of amendments. We can do it again.

Why an Amendment: Today's House Divided

As before the Thirteenth Amendment, our house is divided today: the Constitution protects *your* life, liberty, and property and *my* ability to destroy them through polluting and depleting. As Lincoln found, laws, judicial interpretation, and executive orders can only patch the house temporarily. As in Lincoln's time, only a constitutional amendment can resolve this conflict internal to the Constitution.

Ratifying a constitutional amendment ending pollution and depletion today may seem impossible, but so did ratifying the Thirteenth in 1827, or 1860, or even early 1865. Yet it helped transform this nation's most deadly, divisive issue to one which united us.

I'm not proud of it, but I knew little about abolitionism before working on sustainability. After I learned that Douglass, Lincoln, and others had acted on foundations laid by the anti-slavery authors of the Constitution, I wondered: had anyone laid a foundation opposing pollution and depletion?

I hadn't expected to find any. Humans had barely burned any coal by then. They couldn't imagine global warming, plastic, or carcinogens. They didn't know of extinction.

To my surprise, I found such a precedent. It also came from the framing of the Declaration of Independence and the Constitution, in particular from John Locke's 1689 *Two Treatises on Government*, which are like a blueprint for both founding documents. I wish I'd read them earlier.

An Enlightenment Value We Lost

For context, Locke wrote in England in the early Enlightenment era when Europeans were increasingly questioning monarchy. He wanted to clarify what made governments "legitimate." Why should people choose to submit to another's rule? He concluded that for government to be legitimate and a rational choice, one of its few roles must include to protect its citizens' life, liberty, and property.

How we acquire property was fundamental to his reasoning. If an apple grew in nature, it was no one's property. If you ate it, at some point it became yours. When? He concluded it became yours when you mixed your labor with it: "Whatsoever then he removes out of the state that nature hath provided, and left it in, he hath mixed his labour with, and joined to it something that is his own, and thereby makes it his property."

Locke reasoned that acquiring things by improving nature improved the world. Government could protect property (also life and liberty) better than individuals, so a government that protected life, liberty, and property (plus a few other roles like defense and adjudicating disagreements) improved life. If a government didn't fulfill this role, it was illegitimate. The Declaration and Constitution repeat his language.

But wait. Back to that apple: since others could have picked it if you hadn't, were you depriving them? Locke clarified a limitation on your right to acquire from nature. I found precedent in what Locke wrote next, "For this labour being the unquestionable property of the labourer, no man but he can have a right to what that is once joined to, at least where there is enough, and as good, left in common for others."

You may only acquire from nature if you leave "**at least where there is enough, and as good, left in common for others**"—what I call his "stewardship clause." Locke's clause on legitimate acquisition of property from nature made stewardship fundamental to government's legitimacy.

He clarified that no one is hurt if one person drinks from a river since the river remains full and others can drink their fill too. Land and water, he says, can be claimed "where there is enough of both." As Susan Liebell, 🎙 professor of political science at Saint Joseph's University, wrote, Locke's "theory of property rights forces the individual to justify ownership based upon the needs of the community and the quality and quantity of the resources left for others."[51]

That is, if I drink from a river without depriving you, no problem. But today the Colorado River doesn't reach the sea. Nor do the Tigris, Euphrates, Jordan, or many other rivers. We have claimed water to the point where there isn't enough—or any—for others. Governments including ours allow behavior violating the stewardship clause, which makes them illegitimate, in Locke's view, and not worth submitting to. We can restore legitimacy by restoring stewardship.

After finding Locke's stewardship clause, I found more. Jefferson and Madison considered stewardship in 1789, while discussing the Bill of Rights for the US Constitution. Jefferson considered that if a generation could, in using resources, deprive future generations of them:

> A question of such consequences as not only to merit decision, but place also among the fundamental principles of every government . . . I set out on this ground, which I suppose to be self-evident, "that the earth belongs in usufruct to the living": that the dead have neither powers nor rights over it.

In plain language: "the living have a right only to what they can fairly use

---

51   Susan Liebell, "The Text and Context of 'Enough and as Good': John Locke as the Foundation of an Environmental Liberalism," *Polity* (H-22, Q2) 43, no. 2 (April 2011): 210–241.

during their own lives. To take more than they could repay during their lives would be to rob future generations of resources."[52]

Reflecting on the Constitution's goal to "secure the Blessings of Liberty to ourselves and our Posterity," Jefferson clarified the problem with someone using up resources:

> If he could, he might, during his own life, eat up the usufruct of the lands for several generations to come, and then the lands would belong to the dead rather than the living, which would be the reverse of our principal. What is true of every member of the society individually, is true of them all collectively, since the rights of the whole can be no more than the sum of the rights of the individuals.

In other words, stewardship must hold for all, not some. Stewardship was once an Enlightenment value, but we've lost it. Restoring it is the small adjustment to Pinker's view I alluded to in part 2 that I suspect he would welcome. Locke didn't invent stewardship, of course. Religions have practiced it for millennia and Sustainable Free Abundant cultures yet longer.

Loopholes Neglecting to Protect Life, Liberty, and Property

In New York, you can drink in a bar but not smoke; but you can smoke in your car but not drink. Regarding alcohol and tobacco, for adults we tax selling them but for children we ban selling them. Are these laws inconsistent? No, what makes them consistent is how your behavior affects others.

In a bar, drinking affects only you. If you give yourself cirrhosis, that's your business, but smoking causes cancer and birth defects in other people who can't consent. By contrast, in your car, smoking affects only you. If you give yourself lung cancer, that's your business. But if you drink in your car, nobody consents to their children getting hit.

---

52 "Generational Debt," Thomas Jefferson Foundation.

We *tax* selling adults alcohol and tobacco because adults can consent to their harms, but they cause externalities society must bear, like cigarette butts and beer cans everywhere, though one could argue only the people who litter should pay those taxes.

We *ban* selling them to children because children can't consent. We *ban* smoking in bars because third parties can't consent to getting cancer or being born with birth defects.

A third party getting cancer, being born with a birth defect, being hit by a car, or dying by the millions annually aren't externalities. They are the destruction of life, liberty, and property.

We don't *tax* child labor or slavery. We ban them. Carbon taxes and governments permitting pollution are what logicians call a *category error*. They fail to address the problem. (To be precise, taxing alcohol and tobacco for adults is a category error too. Since cigarette butts don't biodegrade and melting aluminum pollutes, they destroy life, liberty, and property too.)

Beyond failing to solve the problem, these purported solutions augment it. Government gaining revenue from permitting the taking and destruction of what it should protect, motivates it to permit more, growing government and corruption.

Meanwhile, if I poison your food, it's illegal, but putting poisons into the environment that everyone knows will destroy life, liberty, and property is often legal. Buying an airplane ticket that causes millions of times more pollution, therefore destroys more life, liberty, and property, is legal. This discrepency makes sense because until recently, people could reasonably believe pollution went away, but we know otherwise today. This gaping loophole further divides our house.

We are violating more than Locke's stewardship clause. Pollution and depletion destroy life, liberty, and property. I may personally benefit from polluting, but the role of government isn't just to help me; it's to regulate how we affect others.

In Cancer Alley, Flint, Michigan, and Sacrifice Zones—regions so polluted that no one imagines we can rehabilitate them—some people benefited from depriving others of years of their lives, the liberty of living healthily, and the value of their property. In countless ways, from littering to sound and light pollution, our government allows people to destroy others' lives, liberty, and property.

Another loophole: people who get cancer in Cancer Alley face often impossible hurdles like establishing legal standing or proof that their particular cancer came from a particular carcinogen from a particular emitter, yet that emitter can legally release carcinogens everyone knows will destroy life, liberty, and property. Similar loopholes exist in seeking redress from endocrine disruptors, greenhouse gases, and noise pollution. That we allow this destruction of life, liberty, and property makes a mockery of our values of reason, science, and stewardship. Future generations will see us as we see those who allowed slavery.

To be legitimate by the principles of our Constitution, a government must stop activities that destroy life, liberty, and property, including polluting. Some things we can't take from nature and still leave "enough, as good in common for others," particularly fossil fuels, minerals we blow up mountains for, water from aquifers faster than they replenish, and other nonrenewable resources. Extinctions can't be reversed.

It's tempting to say, "But life isn't possible without polluting and depleting." That's our PR-firm-on-steroids riders speaking. Slaveholders convinced themselves that "freedom is not possible without slavery" too.

It's tempting to believe that human ingenuity requires polluting and depleting and we've overall benefited from them despite their harm. If you claim that the greater good sometimes requires sacrifice from others without their consent, you've put yourself in the realm of the Tuskegee Syphilis Study, where doctors infected people with syphilis without their consent, or Josef Mengele, the doctor in Auschwitz who performed horrific experiments on prisoners.

It's tempting to point out that we exhale and poop, call these things pollution, and say life requires polluting. But animals exhaled and pooped for billions of years before humans existed. Behaviors that predate human existence aren't inherently unsustainable (though too high a population density make them unsustainable).

So What's the Problem?

Today, governments not only don't *prevent* the taking or destruction of life, liberty, and property, or of taking from nature without leaving "enough, and as good, left in common for others." They *permit* it, and beyond the sense of allowing it. The US Environmental Protection Agency and bodies acting on the Clean Water Act, for example, *issue permits* to do it, including to generate revenue and profit.

What's the problem with government permitting these activities and what do we call it? Frédéric Bastiat answered these questions in his 1850 book *The Law*. An advocate of Locke's *Treatises*, he wrote that if

> the law takes from some persons that which belongs to them, to give to others what does not belong to them . . . [if] the law performs, for the profit of one citizen, and, to the injury of others, an act that this citizen cannot perform without committing a crime . . . it is not merely an iniquity—it is a fertile source of iniquities, for it invites reprisals; and if you do not take care, the exceptional case will extend, multiply, and become systematic.

He lamented that all "the consequences of such a perversion . . . would require volumes to describe," but one of the main ones is those benefiting feeling entitled:

> No doubt the party benefited will exclaim loudly; he will assert his acquired rights. He will say that the State is bound to protect and encourage his industry; he will plead that it is a good thing for the State to be enriched, that it may spend the more, and thus shower down salaries upon the poor workmen.

Bastiat foresaw politicians today using such arguments promoting permitting legislation such as carbon taxes and dividends, even—especially—business-friendly ones.[53] He warned

> Take care not to listen to this sophistry, for it is just by the system-
> atizing of these arguments that legal plunder becomes systematized.

Plunder, made legal. Is plunder too strong a term? Bastiat "regretted that there is something offensive in the word. I have sought in vain for another," but concluded it was the best and only appropriate word. Government permitting the taking or destruction of life, liberty, and property, or of taking from nature without leaving "enough, and as good, left in common for others" is best described as legalized plunder.

Other results of legal plunder include that those plundered, if they gain power, instead of stopping plunder, tend to plunder back. If the government permits plundering, citizens lose respect for the law.

Uniting Today's House Divided: An Amendment

Though Bastiat wrote of the United States, "There is no country in the world where the law is kept more within its proper domain—which is, to secure to everyone his liberty and his property," he despised slavery as "a violation, sanctioned by law, of the rights of the person." It was legal plunder that abolitionists and anti-slavery politicians took on with laws, judicial interpretations, and executive orders. Lincoln took decades to discover that these tactics wouldn't end the plunderers' entitlement or unite a house divided. He found what did unite it: a constitutional amendment.

Only a constitutional amendment in the style of the Thirteenth Amendment can resolve today's internal contradiction in the most basic role of government in our Constitution. I'm not saying it will solve all our problems, or even all our environmental problems. It may not

---

53   "The Conservative Case for Carbon Dividends," Climate Leadership Council, Baker, James; Paulson, Henry; Feldstein, Martin; Schultz, George; et al.

alone be sufficient, but I believe you will come to agree it is necessary.

Environmental solutions are framed as "big government regulating markets" versus "limited government freeing markets," but what will work is *big* action to *limit* government from legalized plunder.

Just as the Thirteenth resolved the contradiction between protecting freedom and permitting slavery, I propose an amendment to clarify what the framers would have written if they could have foreseen the results of letting stewardship lapse—that is, of a government allowing people to deprive others of life, liberty, and property without their consent when that deprivation is mediated through the environment.

I envision it containing three clauses:

> Within the United States, or any place subject to their jurisdiction, nothing may be taken from nature except where there is enough, and as good, left in common for others.

> Within the United States, or any place subject to their jurisdiction, nothing may be taken from or put into nature when doing so destroys others' life, liberty, and property.

> Congress shall have power to enforce this article by appropriate legislation.

One might call this Anti-Plunder, Protecting Life, Liberty, and Property, Imperialism-Ending amendment an APPLL PIE Amendment. If you'll indulge me spelling it as I pronounce it, I call it an APPLE PIE Amendment.

Is the APPLE PIE Amendment Possible?

I asked you about your deepest values because such an amendment would deliver on them, no matter how skeptical you may feel at first. That skepticism is your PR-firm-on-steroids rider speaking.

This amendment is about more than pollution and depletion. It's about our deepest values and what constitutes citizenship. Freedom, equality, and democracy can only exist in the long term with stewardship and without plundering, we just didn't notice while we believed resources were infinite. Stewardship is as important a value as these other values. For 250,000 years, people lived close enough to nature that stewardship was automatic, as it still appears to be for many indigenous cultures today. When humanity began to live unsustainably but impacted only a small fraction of the earth, few people noticed, if any. Now we live when humanity's impact is lowering earth's ability to sustain life. Without stewardship, predictions that billions of people could die from PAID-culture practices within our lifetimes are credible.

What nearly no one gets and I wouldn't have believed until my experiences revealed it, is that practicing stewardship will improve your life. It doesn't mean deprivation and sacrifice, nor risk of returning to the Stone Age. It doesn't require willpower. We will wish we had restored it earlier. We will look back at PAID culture with the repugnance we look at slavery and former addicts who look back at destroying their friendships, families, and lives.

This amendment will free markets from their coercion and corruption we see today. No market based on pollution, depletion, plunder, or addiction is free. It will liberate innovators to solve problems. It will rally everyone to support it as everyone contributed to the war effort in WWII, like growing victory gardens, buying war bonds, and using less oil at home. Only no one will have to take up arms for this cause.

The more I learned of the Thirteenth Amendment, abolitionism, anti-slavery politics, Enlightenment values, indigenous cultures, and the harms of pollution and depletion the more the APPLE PIE Amendment made sense. Soon I realized it was essential and came to the conclusion that an overwhelming majority of Americans would come to agree. Today it's as hard for us to imagine it working as the Thirteenth would have seemed in 1827.

If you haven't gone through your mindset shift, you may find it hard to believe. Whatever your deepest values, you will find it increasingly meeting them.

Is It Prudent?

As long as slavery exists in a market, that market is coercive, not free. If a market including slavery grows, it grows coercion, not freedom. Before the Thirteenth Amendment, people advocated limited slavery and gradual abolition. The Thirteenth Amendment ended it immediately. People may have considered ending slavery immediately "extreme," but they didn't understand that markets exploit every loophole they can (as with the clause "except as a punishment for crime whereof the party shall have been duly convicted"). If we permit plunder, and we convince ourselves that that coercive market is free, then *that market will grow coercion and tyranny*. It will divert resources from innovators who solve problems best to people who coerce the most. As Friedman saw, "The greatest source of inequality has been special privileges granted by government," such as the special privilege to claim legal ownership of another person.

The solution to slavery's coercion and tyranny wasn't to slow it with taxes, which would grow government and create an incentive to promote more slavery. Nor would developing technology solve slavery, as efficiency accelerates and grows the culture wielding it, as Whitney's cotton gin did. Nor would solving piecemeal the problems it created, like enabling separated families to talk by some past equivalent of social media or developing medicine to heal whip lashes. Relying on the market doesn't work either. Slavery was declining in the late eighteenth-century United States. Many predicted it would end on its own. The cotton gin—technology—revived it.

Technology, markets, and taxes don't remove the coercion and tyranny. Applying them is a category error.

What stops tyranny and coercion? Stopping tyranny and coercion does. Letting people trade freely will grow freedom, not tyranny.

Advantages of an **APPLE PIE** Amendment

It's Consistent With Existing Law

It bears repeating: When we learned second-hand smoke caused cancer we could have simply increased taxes on smoking indoors, but we didn't. Cancer isn't an externality. We ban selling alcohol and tobacco to children, drunk driving, child labor, and slavery. We don't tax them. It goes without saying that polluting and depleting kill people and those affected can't consent to it. How do you consent to being born with a birth defect, for example?

We don't limit theft by taxing it or creating cap-and-trade schemes for it, even theft that only deprives people of material property that can easily be restored and doesn't affect the victim's life or liberty. We don't limit assault through taxes or cap-and-trade schemes either, even when the injury isn't permanent and we know the victim's body will heal to the point where there will be no sign of the injury. We know that permitting theft and assault would undermine the basis of society and civilization.

If what you create, I can take or destroy with impunity, you lose sight of a better future. You lose incentive to work or improve the world. We don't benefit from living under a government when it allows theft or assault so we don't allow those behaviors. Yet polluting, depleting, and plunder hurt people more than assault and deprive people of more than theft. Making polluting, depleting, and plunder illegal is consistent with other behaviors we make illegal, like murder, rape, slavery, and arson.

We only recently changed laws to ban indoor smoking and drunk driving. In our defense, we didn't have reason to ban second-hand smoking before we knew how deadly it was. We didn't say we should balance the needs of people who benefit from it with those it gave cancer to. People have gotten drunk since before civilization, but if they rode a horse drunk, they weren't more dangerous. It took time to realize how deadly drunk driving was to everyone, not just the driver, but when we learned, we responded to evidence and banned it. We didn't create cap-and-trade schemes to slowly limit these behaviors. We don't consider the deaths they cause "externalities."

Like drunk driving and second-hand smoke, we only recently learned how much polluting and depleting harm and kill. It makes sense that until now we thought to treat them like smoking outdoors or riding a horse drunk, but we know differently today. It follows precedent to change how we treat them as we changed how we treat smoking indoors, selling children cigarettes, and drunk driving.

It Achieves Ecomodern and Techno-optimist Goals

I love innovation and technology, hence my patents, satellite, the companies I started, my PhD in physics, and my MBA. It's tempting to think the APPLE PIE Amendment would inhibit technological innovation. PAID culture teaches that human ingenuity can improve the world but needs energy: "Cheap, plentiful, reliable energy we get from fossil fuels and other forms of cheap, plentiful, reliable energy, combined with human ingenuity, gives us the ability to transform the world around us into a place that is far safer from any health hazards (man-made or natural), far safer from any climate change (man-made or natural), and far richer in resources."[54] If any problems arise, "we can use fossil fuels to help solve fossil fuel problems—to transform waste from a more dangerous form to a less dangerous form, or even to a benefit, by using energy and ingenuity."

From that perspective, avoiding using fossil fuels or any other energy sources sounds inhibiting, but the APPLE PIE Amendment doesn't disallow burning fossil fuels, just how you affect others' life, liberty, and property. If you can extract from nature so the reserves remain and release nothing that destroys life, liberty, or property, you can do it. I hope human ingenuity finds ways to do it.

Flying currently requires destroying life, liberty, and property and plundering, but suppose it could be done otherwise. The worst way to

---

54   Alex Epstein, *The Moral Case for Fossil Fuels* (2014).

get there is to allow flying as practiced today, which keeps the market coercive, not free. If we can find ways to fly without coercion, the fastest, most effective way there is to stop flying that requires it.

The APPLE PIE Amendment promotes innovation that doesn't destroy life, liberty, and property. It levels the playing field. It enables allocating resources to people who can solve problems best. It diverts resources from people who coerce and tyrannize.

If you promote a circular economy, decoupling, dematerialization, substitution, free trade, free markets, green growth, ecotourism, and such, you believe in what the APPLE PIE Amendment will achieve. If you don't believe so, whatever economy you describe is not circular; you haven't decoupled, your growth isn't green, etc. If you don't believe the APPLE PIE Amendment would work, you don't believe those ideas you promote work.

It Limits Government and Frees Markets

The APPLE PIE Amendment rejects and protects from central planning. It will shrink government since so much of government's growth is funded by fossil fuel and other extractive interests. Consider the views of a few prominent proponents of limited government beyond Bastiat.

Often called the founder of conservatism, Edmund Burke wrote on respecting tradition instead of jettisoning values like stewardship and warning of the dangers without it. "One of the first and most leading principles" for laws is to avoid that people

> unmindful of what they have received from their ancestors, or of what is due to their posterity, should act as if they were the entire masters; that they should not think it among their rights to cut off the entail, or commit waste on the inheritance, by destroying at their pleasure the whole original fabric of their society; hazarding to leave to those who come after them a ruin instead of a habitation.[55]

---

55 Edmund Burke, *Reflections on the Revolution in France* (1790).

I expect future generations will see humanity's unsustainability as violating long-established tradition. I expect current generations will too.

Russell Kirk's landmark *The Conservative Mind*'s comment on that passage reinforcing its relevance today:

> If men are discharged of reverence for ancient usage, they will treat this world, almost certainly, as if it were their private property, to be consumed for their sensual gratification; and thus they will destroy in their lust for enjoyment the property of future generations, of their own contemporaries, and indeed their very own capital . . . The modern spectacle of vanished forests and eroded lands, wasted petroleum and ruthless mining, national debts, recklessly increased until they are repudiated, and continual revision of positive law, is evidence of what an age without veneration does to itself and its successors.

Kirk and Burke identified "leveling instincts" as major threats to their values. They mostly missed that unsustainability, technologies based on it, and the pace of change it produced threatened them more, despite their rise often coming from people who intended to support those values.

Putting the results of the APPLE PIE Amendment in their language: it restores virtue, morality, prudence, ethics, duty, integrity, and tradition, among others of their values. It opposes sloth, greed, gluttony, and the like. It leads to glory, honor, grace, community, and family.

Nobel laureate in economics Friedrich Hayek wrote in *The Road to Serfdom*, "The harmful effects of deforestation or of the smoke of factories cannot be confined to the owner of the property in question . . . these tasks provide a wide and unquestioned field for state activity." He continued, "We shall never prevent the abuse of power if we are not prepared to limit power in a way which occasionally may prevent its use for desirable purposes."

Hayek wrote those words watching Hitler, Stalin, and Mussolini rampaging across Europe, destroying freedom. He promoted freedom. If he were alive today and saw pollution and depletion destroying life, liberty, property, and freedom to today's extent, I believe he would analyze PAID culture as he did central planning and find similar patterns. Central planning is different from polluting and depleting, but both coerce people and lead central planners as well as polluters and depleters to coerce ever more to maintain their world view. Both also arise from people genuinely wanting to help others, just not realizing where their path leads.

Milton Friedman wrote, "There is a real function for government in respect to pollution: to set conditions and, in particular, define property rights to make sure that the costs are borne by the parties responsible." Also, "The role of government just considered is to do something that the market cannot do for itself, namely, to determine, arbitrate, and enforce the rules of the game."

The equality the APPLE PIE Amendment promotes is for equal protection not of outcome but of protection of life, liberty, property, and therefore freedom. That is, it's tempting to say that people who innovate and produce may stratify society, which may look like haves and have-nots, but they improve it so that even the poor enjoy benefits beyond what they would otherwise. That claim misses my point. I don't object to hierarchy in general, but to dominance hierarchies that depend on and require destroying life, liberty, and property. The APPLE PIE Amendment allows you to produce all you want, but to accept that government will protect others from your destroying their life, liberty, property, and freedom without their consent.

Barry Goldwater said, "I am a great believer in the free, competitive enterprise system and all that it entails, I am an even stronger believer in the right of our people to live in a clean and pollution-free environment."[56] He continued, "When pollution is found, it should be halted at the source, even if this requires stringent government action against important segments of our national economy." "Our job," he said, "is

---

56 Daniel Farber, "The Conservative as Environmentalist: From Goldwater and the Early Reagan to the 21st Century," *Arizona Law Review* 59 (2017).

to prevent that lush orb known as earth . . . from turning into a bleak and barren, dirty brown planet." He also said, "The administration is absolutely correct in cracking down on companies and corporations and municipalities that continue to pollute the nation's air and water." He also said, "Extremism in the defense of liberty is no vice."

William F. Buckley said on government regulating pollution: "Here is a legitimate concern of government—a classic example of the kind of thing that government should do, according to Lincoln's test, because the people cannot do it as well or better themselves."[57]

Ronald Reagan said, "I'm proud of having been one of the first to recognize that states and the federal government have a duty to protect our natural resources from the damaging effects of pollution."[58]

Conservative, libertarian, and classical liberal voices embraced stewardship and government protecting its citizens from deprivation of their life, liberty, property, and freedom. The difference between those I just quoted and their counterparts today isn't that the values of today's group changed, as far as I can tell. They benefited from our government legalizing plunder.

They benefited from and participated in behavior violating their values including government growing from revenues from polluting, depleting, and plundering. They benefited from collusion between government and plundering industries. They benefited from taxes from those practices and more. Like me when my first company, Submedia, promoted soda to kids and more, they allowed their values to be corrupted.

They've violated their values and profited from it enough to create PR-firm-on-steroids riders telling them for decades that that coercive, deadly markets were free, that their wealth came only from helping people (some did, but much from coercion), and that their actions were good, right, and normal when they were actually coercive. I know from experience

---

57 Farber, "The Conservative as Environmentalist."

58 Farber, "The Conservative as Environmentalist."

how tempting it is to succumb to that voice and say I was helping riders and so on. We all do. It's still corruption of our values.

Hayek was clear on collusion and coercion: "Once wide coercive powers are given to governmental agencies for particular purposes, such powers cannot be effectively controlled by democratic assemblies." We enjoy the benefits that polluting and depleting allow, but he continued, "If we wish to preserve a free society, it is essential that we recognize that the desirability of a particular object is not sufficient justification for the use of coercion."

It Reduces Corruption

Permitting destruction of life, liberty, and property is corruption, leading to coercive markets, further resulting in infiltration of government, corruption, and regulatory capture. McGill University researchers found: "The oil and gas and mining industries are the fourth and fifth most corrupt industries globally," with the first two being related: "public works and construction [and] utilities."[59]

It Is Anti-Racist and Anti-Imperialist

Racism resulted from slavery, which results from imperialism, which resulted from living unsustainably. If we could magically end all racism everywhere, but kept living unsustainably, we'd recreate the conditions that caused it. It's tempting to see racism and similar results of dominance hierarchies affecting people here and now and say we have to solve the immediate problem first, suggesting our environmental problems are too much to handle for people struggling to make ends meet. While I agree people of higher status must take the most responsibility, unsustainability lies upstream of racism.

Future generations will see humanity's foray into unsustainability as

---

59   Isabelle Côté and Dalia Feldberg, "The Demand and Supply Sides of Corruption in the Extractive Industries, Including Parliamentary Oversight of Mining Companies/Activities in Home and Host Countries," McGill: Parliamentary Programs, Research, and Publications, May 20, 2022.

driving imperialism, racism, and social injustice. I expect current generations will too.

If I put the results of the APPLE PIE Amendment in their language, it restores caring, fairness, and social justice, among others of their values. It opposes racism, sexism, and similar patterns of domination. It leads to caring, fairness, mutual support, justice, social justice, community, and family, not selfishness.

It Promotes Freedom, Equality, and Democracy

With this amendment, you are free to do what you want, but government will play its role in stopping you from destroying others' life, liberty, and property and from plundering. This amendment creates the liberty that traffic lights and laws banning smoking indoors do.

People benefiting from destroying others' life, liberty, and property and plundering may not like giving up their benefit, as Locke and Bastiat foresaw. People who benefited from slavery didn't want to give up their benefits, but if you believe a government protecting life, liberty, and property benefits all of us, you see that even slaveholders benefited overall from abolition. The Thirteenth Amendment increased freedom, no matter how much slaveholders feared otherwise. The APPLE PIE Amendment will increase freedom and liberty for all too, even those with the most status today.

Enforcement Is Easy

An amendment is only useful if it can be enforced. Enforcing an APPLE PIE Amendment will be straightforward for meaningful infractions.

Historical precedent shows the importance of enforcement. Enforcement is why Britain succeeded in abolishing its slave trade while Reconstruction ultimately failed, and why Prohibition failed. It was widely predicted that if Britain stopped trading slaves, other slave-trading nations would fill the gap, resulting in no decrease in overall trade, just lost profits for Britain. But Britain had financial interest for others to follow and the world's most powerful navy, which it used to prevent the slave trade, which led other slave-trading nations to ban it too. By contrast, the US government enforced the Reconstruction amendments only temporarily. When enforced, Reconstruction worked. When enforcement stopped, it failed. Prohibition, where anyone with grains and a bathtub could distill alcohol, was impossible to enforce. It failed.

The locations of deposits of most fossil fuels, minerals, and fissile materials are well-known, so we know where to monitor for attempts to extract them. Extraction requires hundreds of millions of dollars and years of development. Most activity like mining sends seismic signals detectable for large distances. You can't hide the biggest applications of fossil fuels or use of fission either, like container ships, jets, coal plants, nuclear reactors, or oil refineries. Carcinogens and poisons tend to have specific applications where it's well-known where they'll be applied, like industrial agriculture. Likewise with uranium and other fissile material.

Other large-scale forms of pollution, depletion, and plunder are hard to hide, such as factory farms and industrial agriculture that depletes aquifers and rivers. We know where the aquifers and rivers are.

If the only effect of the amendment were to end these easy-to-find large-scale uses, downstream pollution would decrease too. Smaller-scale pollution would then decrease too, since most of it depends on fossil fuels, like plastic and artificial fertilizers,

It Increases National Security

Americans often talk about the problems of "dependence on foreign oil." They treat "foreign" as the key word, but "dependence" undermines national security. Beyond dependence on oil, depending on a power grid that could fail or be attacked makes our security more vulnerable and brittle. If a large number of households gained my resilience to grid power going down, the power grid would not require 99.9 percent uptime. We could shrink it, which would lower risks of terrorist attack and blackouts as well as taxes and monthly bills.

Besides, we already let our military destroy life, liberty, and property. Nearly everyone accepts common defense as a legitimate role of government. If we let our military pollute and deplete as well, it would have plenty of fuel without the rest of us consuming it.

A nation that accepts Cancer Alley, Sacrifice Zones, and other health hazards is not protecting its citizens' security. The APPLE PIE Amendment would end this legal plunder and stop new cases from forming.

It Promotes Health, Family, and Longevity

Pollution destroys health and shortens life spans.

It's tempting to think of planes as connecting family. *One* plane may bring us to a distant family member. *Flying in general* leads us to live flying-distance from family members, leading to less time with family.

Without flying, we would spend more time with family, if desired, not less. People in flying-distance families complain as if their problems were universal, but they are a privileged minority. According to *The Atlantic*: "Straying from family is unusual in the US: roughly three in four American adults live within 30 miles of their nearest parent or adult child, according to a 2019 study. Only about 7 percent have their nearest such relative 500 or more miles away."[60] Pew research shows, "Having extended family nearby also differs by income status. Adults with lower and middle incomes are more likely than upper-income adults to live near at least some extended family. In contrast, upper-income adults are the most likely to say they live near no extended family."[61]

Hospitals might not have stents and end-of-life treatments that promote lives by months to years, but markets freed from coercion would unleash innovators to create new solutions. Also, we wouldn't have doof or pollution. Humanity's greatest health advances either don't require pollution or

---

60 Stephanie H. Murray, "How Affluence Pulls People Away from Their Families," *Atlantic*, May 11, 2022.

61 Kiley Hurst, "More Than Half of Americans Live within an Hour of Extended Family," Pew Research Center, May 18, 2022.

depletion, like soap and hygiene, or date back centuries or millennia, like anesthesia, antibiotics, and vaccines. We don't have to give them up and can keep developing them. The amendment would stop us sacrificing billions to save tens of thousands.

It Unites Across Political Divides

Seeing why abolition unites Americans today shows how sustainability will too. The Thirteenth Amendment transformed the most divisive issue in American history into one of unity. Politicians of nearly all political leanings claim Lincoln, Douglass, and other anti-slavery politicians as their own.

Slavery didn't only divide north from south. Much of the business community supported slavery, stating in 1861, "No saner or more just agreement was ever made working to the mutual benefit of both."[62] But slavery violated the business community's values. Businesspeople supported it not out of principle, but from being corrupted by profits from coercive markets.

In this book, I tend to quote limited government, free trade economists and politicians more, partly because big government proponents already tend to support government action to protect those most hurt by pollution, depletion, and plunder. To clarify, by applying limited government, free trade values and beliefs relevant to pollution, depletion, and plunder, I am not suggesting their ideas apply elsewhere or not. I quote them because the action necessary for government to take is big, but, as with the Thirteenth Amendment, in the direction of limiting government from permitting plunder, and making trade more free, less coercive. After business exited slavery's thrall, Milton Friedman called slavery, "A disgrace to this country."

Since he knew, "Businessmen favor free enterprise in general but are opposed to it when it comes to themselves," I suspect he disliked the Thirteenth Amendment's loophole allowing slavery "as a punishment for crime whereof the party shall have been duly convicted." Since Reconstruction,

---

62 William Sumner Jenkins, *Proslavery Thought in the Old South* (1935).

businesses used that exception to deprive African-Americans of freedom and more, which persists in today's prison system.

People of all political stripes will celebrate the APPLE PIE Amendment and claim it as theirs, as they do with the Thirteenth.

Today's Libertarian Party's 2023 platform statement that says, "Laws should be limited in their application to violations of the rights of others through force," would seem to oppose destroying others' life, liberty, and property through pollution and depletion as much as by slavery. [63] Likewise, across the political spectrum with France's Declaration of the Rights of Man and of the Citizen: "Liberty consists in the freedom to do everything which injures no one else . . . The Law has the right to forbid only those actions that are injurious to society."

It Uniquely Addresses the Problem Directly

Many proposed solutions approach the problem from the side. If the problem is that polluting and depleting hurt people, this amendment directly addresses that problem, just as the Thirteenth Amendment did. Innovation, market incentives, and legislation don't directly stop the problem, let alone vague ideas like "overthrowing capitalism."

Objections

"It can't pass." Neither could the Thirteenth until it did. Nor could people walk on the moon or run a four-minute mile.

"The Thirteenth only passed because of incredible circumstances that will never repeat." Generations of people acting created those circumstance. Passing the APPLE PIE Amendment requires a movement creating overwhelming popular support first. I don't propose starting by marching to Capitol Hill. I propose starting top-down, bottom-up, everywhere all at once starting here now with you and me, using the vision of this amendment to create meaning, purpose, direction, and enthusiasm.

---

63 "Platform," Libertarian Party, May 2022.

"If we don't pollute, others will; they'll beat us in the market and militarily." People objected this way to abolitionism, banning smoking indoors, stopping selling cigarettes to children, and countless other examples and it was wrong there too. Much of China's polluting, depleting, and plundering happens for the purpose of export. Much of Russia's strength comes from fossil fuels. Many nations are already more sustainable than the US (a low bar). Assuming they implement similar solutions, China and Russia will lose revenue and their citizens will see the growing gap in health and safety. They will follow as all the slave-trading empires followed England to ban slavery and New Jersey followed New York to ban smoking indoors.

If you fear coercive markets will beat free markets and you disagree with John Locke, Adam Smith, and traditions that followed them, maybe you prefer moving to Russia or China. But I stand with Abraham Lincoln, who saw that plantation owners included the wealthiest people in the world, but he knew that freedom and free markets beat coercive, plundering markets.

Likewise, if you believe imperialism, autocracy, and fascism can beat democracy and equality before the law, maybe you'd prefer living under them instead of a more democratic, free nation.

"It won't stop all polluting, depleting, and plundering." If it only stopped the biggest cases, it would still be more effective than anything that now exists to restore government that protects life, liberty, and property. See the section above on enforcement.

The lack of the APPLE PIE Amendment is the greatest impediment to solving today's problems of pollution, depletion, and plundering.

"It will fail like Prohibition." Unlike Prohibition, it isn't legislating what people do to their own bodies. It is regulating people harming others. It isn't regulating a practice going back thousands of years. Pollution barely existed more than two hundred years ago.

Also unlike Prohibition, it will be easy to enforce, as noted above.

"It's too vague." The Constitution doesn't define free speech or slavery. Most amendments are written broadly, to allow interpretation.

"**We need time to transition.**" It's tempting to say we should build clean, green, renewable alternative energy sources first, then lower our use of polluting, depleting energy. Sounds great, but no alternatives exist for many things PAID culture depends on, like oil, flying, long-distance trucking and shipping, steel, and microchips. Also existing alternatives aren't clean, green, or renewable, as shown by the resources in the coda. Also, when we introduce new energy sources, we don't use them to replace old ones; we use the old ones *and* the new ones.

Lincoln first promoted gradual abolition but changed his mind. Extra time to transition may have made sense from the slaveholder's view, but not from the slaves' view. Would *you* say to them, "We believe it's wrong to keep you enslaved, but the people enslaving you don't want to transition too suddenly. Would you mind remaining enslaved for them?" The Thirteenth Amendment went into effect immediately.

"**It has no precedents.**" In addition to the precedent of the Thirteenth Amendment, Senator Gaylord Nelson, the father of Earth Day, proposed a constitutional amendment guaranteeing the right to a clean environment in 1970.

Maya Van Rossum 🎤 is leading the movement for an amendment like Nelson proposed—what she calls a "green amendment"—starting with states.[64] Three states have green amendments with twenty states developing theirs (and counting). Green amendments and APPLE PIE amendments will complement and augment each other.

"**States should act first.**" I agree. Three states already have green amendments with another twenty developing theirs. States abolished slavery before the nation did. The APPLE PIE Amendment will start with states too.

"**Life today requires things that pollute, like flying and disposable diapers.**" Democracy in the Antebellum South "required" slavery too, or so slaveholders tried to convince themselves.

---

64 "Green Amendments for the Generations."

Ultimately, business leaders must face the choice that plantation owners after the Thirteenth Amendment and German businesspeople who bet on Hitler's growth and expansion did: if their business models require coercion and death, they can choose to stop, as Carter and Schindler did.

Political and other leaders have the opportunity to make the choices that Lincoln, Bonhoeffer, Wilberforce, Clarkson, Wallenberg, and Sousa Mendez did. We consumers of coercive, deadly businesses have to make the choices that people who relied on slave-produced cotton, sugar, and so on did, or the choice made by those who watched Kristalnacht but kept paying their taxes.

The longer we procrastinate, the harder the choices become, but the alternative is to live with temporary material comfort replacing meaning and purpose—that comfort sustained by internal conflict and external suffering and death. It would ultimately lead to our equivalent of a Civil War or World War II, which resulted from procrastination in the face of coercion and death.

"It will take too long to pass." Ninety years passed from Pennsylvania's first abolitionist society to ratifying the Thirteenth Amendment, which was too long. Passing the Nineteenth Amendment, for women's right to vote, took about a century, also too long. All the more reason to start now.

By contrast, a gay politician on my podcast, Eric Bottcher, 🎤 told me of meetings for LGBTQ rights where participants planned political action. They dreamed of what they could achieve in fifty years if they were lucky. Two years later, they achieved beyond those dreams.

"My family lives flying-distance away. My work requires flying. Where I live requires air-conditioning and driving." Plantation owners set up their lives to require slavery too. Should we have stopped abolitionism for their dependence on slavery? We made mistakes. The way to solve these problems is not to multiply them.

Ultimately we whose lifestyles require plunder must face the choices that people who lived on plantations or whose livelihoods depended on slavery did. Plantation owners didn't want to perform manual labor.

Cotton and sugar consumers didn't want to know where their goods came from. Ending slavery required them to face and overcome these cruel dependencies. We must face and overcome ours if we want to live by our values. Once we adjust to not living flying-distance from families and so on, we'll wish we had changed earlier. You tell me what you fear losing, and I'll tell you what you'll gain.

"We'll return to the Stone Age or devolve into a Mad Max apocalypse." Slaveholders predicted as much without slavery. They were wrong.

Anthropology has debunked the linear model of cultural evolution underlying this objection. Cultures don't only go forward or backward. Many examples exist of cultures living sustainably for tens of thousands of years and more.

"Rich nations transitioned from dirty to clean. We had our demographic transition. Poorer nations deserve to transition too. We have no right to deny it from them." We aren't cleaner than ever, we mostly moved our pollution and depletion overseas. Much of China's pollution comes from manufacturing products we used to manufacture in the US. China and other nations manufacture cheaper in part because they have laxer protections. Thus we have caused pollution to increase, wrongly claiming innocence.

We can *improve* our quality of life by lowering our environmental impact to below those of poor nations. Then we can help others through a process to improve their lives by reaching our lower levels too.

"People don't want to change. Only making sustainability easier and cheaper can make them change." Before a person's mindset shift, the choice appears between comfort and struggle. The shift reveals that the choice to struggle creates meaning and freedom. Then we see the choice as between procrastinating and engaging with purposeful effort. The Spodek Method can create a mindset shift through a short conversation. Do it with a few people you spend time with and the process of continual improvement feels like swimming downstream.

As for people not wanting to change, if you only consider us here, you miss the people affected by our actions. In 1860, few voters wanted to end slavery either. The slaves did, though.

"It will create a nanny state." Unlike Prohibition, polluting, depleting, and plundering destroy *other people*'s life, liberty, and property, so the APPLE PIE Amendment isn't nanny-state regulation. Polluting, depleting, and plundering are more like secondhand smoke giving others cancer.

"We have to reach Mars. An asteroid may destroy earth. Therefore we must not restrict innovation, even if it pollutes." Our current ways of pursuing space travel, in which governments sanction private individuals and companies, are the opposite of free because rockets deplete, pollute, and plunder heavily, destroying life, liberty, and property. If you believe free markets solve problems optimally, continuing this way delays solutions because it is coercive. More immediately, it promotes corruption, the path to tyranny.

The fastest, most effective way to reach space travel, if possible, is not through coercion.

Advantages

Before considering the advantages this amendment would bring, let's recall the costs of the culture it would end.

There's no denying that profit from destroying or taking others' life, liberty, and property as well as plundering can be used to create comforts, conveniences, and capabilities. It's tempting to think only of losing them if we take them for granted or ignore where they came from. If you look only at the parts you like of anything, it will look good. Plantation homes were beautiful if you ignore that they were built on slavery. The United States' westward expansion looks wonderful if you considered North America empty or populated only by "savages" who benefited from colonists "improving" it.

Like those examples, polluting, depleting, and plundering come at a cost, even if that cost was invisible for centuries. Today those activities are killing people by the tens of millions, projected to grow to billions.

This amendment will create many desirable results, for example:

- Government protecting life, liberty, property and protecting against legal plunder.
- Greater health, safety, personal security, and national security.
- Reduced government size and corruption.
- Reduced imperialism, colonialism, slavery, and racism.
- Reduced coercive markets; level their playing fields.
- More innovation and technologies that improve lives.
- Closer families and communities, less isolation.
- Greater civic participation, arts, culture, sports, and exercise.
- Internal conflict and death spirals replaced with integrity and continual improvement.
- Less addiction and doof.
- Healthier food, soil, and ecosystems.
- Greater global cultural diversity, including small-town America.
- Stabilized population levels at sustainable levels.
- Restoring the Golden Rule, stewardship, and common decency.
- Less litter and garbage.

Do these benefits sound too good to be true? They do if you haven't tried living sustainably. After your mindset shift, the more you continually improve, the more obvious you will see them.

How Impossible the Thirteenth Amendment Looked. Yet It Passed.

I grew up believing abolitionism resulted inevitably from progress, but it was not inevitable. Even most northerners opposed it, many violently. They rioted against it, destroyed property of abolitionists, and even murdered some. Slaveholders held more slaves, more power, and were growing more than ever in 1860. It's hard for us to imagine how much Americans opposed rights for African-Americans including voting, serving on juries, and serving in the military, even many abolitionists. Many believed racial equality would destroy the nation.

William Lloyd Garrison considered the Constitution pro-slavery. The Fugitive Slave and Three-Fifth clauses led him to call it "a Covenant with Death, an Agreement with Hell" and burn a copy of it.

Frederick Douglass for a time read it that way, but by his "What to the Slave Is the Fourth of July?" speech, he read it differently. Then he called it a "Glorious Liberty Document," enabling an anti-slavery movement.

Who was right? Historians today agree with Douglass. They read that the anti-slavery framers of the Constitution fought to restrict slavery to states where it existed, making freedom national and slavery local. They allowed no mention of people as property. As historian Sean Wilentz wrote, "The framers left room for political efforts aimed at slavery's restriction, and eventually, its destruction, even under a Constitution that safeguarded slavery."[65]

That reading enabled Douglass, Lincoln, and others to form an anti-slavery movement to make slavery illegal everywhere (with a loophole for convictions). It took decades and a Civil War, but they succeeded with the Thirteenth Amendment, an outcome inconceivable when proposed or even a year before.

Generations of people who could have owned slaves did what they could. They created anti-slavery groups here, protested there, and acted as best they could. They probably felt hopeless. Peers told them what they did didn't matter. Opponents accused them of destroying democracy. Still, they acted. As historian Eric Foner wrote, "As the *Chicago Tribune* noted at the end of the Civil War, in crisis situations beliefs once 'pronounced impractical radicalism' suddenly become 'practical statesmanship.'"[66]

However hopeless our situation today looks, we may recall that, as

---

65 Sean Wilentz, *No Property in Man* (2018).

66 Eric Foner, *The Fiery Trial* (2010).

Foner continued, "As late as 1858, the *Chicago Tribune*, a strong voice of antislavery radicalism, stated flatly that 'no man living' would witness the death of American slavery." Yet it happened within a decade.

A sustainability movement could achieve as much.

Other Mobilizations

On the topic of stewardship mobilizations, I'll mention a few that would complement the APPLE PIE Amendment, though treating them fully would go beyond this book's scope.

Stopping deliberately addicting citizens: Recalling that addiction doctor Carl Erik Fisher called addiction "a terrifying breakdown of reason," deliberately causing addiction undermines Enlightenment values, indigenous wisdom, and the US Constitution. Laws prevent coercion through threats, pain, and violence. Laws prevent addicting children too young to consent. Laws say contracts are not valid if signed under coercion.

If marketers have developed ways to control the parts of our motivational systems that cause addiction more effectively than we can withhold consent, are they not coercing us, making markets unfree? If so, banning deliberate efforts to act on the brain's mechanism that results in addiction seems in the interests of reason, freedom, and democracy.

National service: There have been many calls for a year or two of national service from all citizens.[67] Sustainability will require rebuilding infrastructure and agriculture nationwide and globally. It will also mean stopping using power requiring polluting, depleting, or plundering, meaning more hands-on work. A year or two of national civilian service would aid transitioning to sustainability. I envision much of it focused on regenerative agriculture and building new, sustainable infrastructure.

A civilian service academy: I've had the honor to have helped lead leadership workshops at West Point with General Lloyd Austin, now Secretary of Defense. The US service academies are among the premier

---

67 Stanley McChrystal, "Every American Should Serve for One Year," *TIME*, June 20, 2017.

institutions for developing leaders, though focused on military service. Should millions of young people perform national service, I envision a service academy with all the rigor of our military service academies, but teaching civilian service and leadership in trades.

CHAPTER 19

A BRIGHTER FUTURE: WHAT IF EVERYONE ACTED SUSTAINABLY?

As winning sports coaches say, when you do your best and focus on the basics, the score takes care of itself. Sustainability and stewardship alone will not create peace on earth. Freedom and democracy require work. People will still conflict with each other on tax rates, abortion, and countless other issues. A culture of sustainability just means we won't sleepwalk toward tyranny and billions of innocent people suffering.

Future generations will see the APPLE PIE Amendment as we see the Thirteenth. It will unite us. They will wish we had passed it earlier. They will think it was inevitable and wonder why we took so long. They will look back with horror at how long we procrastinated. Their history books will honor those who helped make it happen.

We will see environmental laws how we see traffic laws today. Even when in a hurry but stopped at a light, we know they benefit us. We're glad that red means stop everywhere in the world. We fear driving in countries that don't enforce traffic laws.

I'll share what I consider a likely path if we don't embrace stewardship. Then I'll share what I advocate: the fastest, most effective path possible to the healthiest, safest outcome possible.

The Procrastination Option

You've read predictions of possible catastrophic environmental outcomes. I won't detail them since before a mindset shift, many people interpret them as doom and gloom, which leads them to disengage, but peer-reviewed articles go as far as suggesting billions of people could die from environmental problems within a century.[68]

Suffice it to say, everyone could suffer, including the wealthiest. We imagine that when we feel the problem we'll act, but in a global system with feedback and delays, that's like saying we'll hit the brakes when the bumper touches the brick wall at high speed. Besides, those impacting the environment most will feel the problems last, so they will be the last to change.

The Purposeful Effort Option

Let's consider what could happen following a movement restoring stewardship ending plunder.

For a few months, little would seem different as small numbers of people go through their mindset shifts, but conversations on environment and culture will take a new tone. Instead of blame, complaints, and abstract debate, we'll share the freedom, joy, and fun we find in stewardship. As more people experience mindset shifts and learn to lead others through them, more people will lead others through processes of continual improvement.

Then the tide will turn, as enough people practice stewardship that it becomes the norm again after an absence of hundreds of years. Hundreds of millions of Americans will improve the quality of their lives, health, safety, and security by reducing their consumption and impact 90 percent or more within a few years. Other nations will participate and catch up. Many will outdo us. The nation will feel mobilized, energized, and unified.

---

68   Joshua M. Pearce, "Quantifying Global Greenhouse Gas Emissions in Human Deaths to Guide Energy Policy," *Energies*, no. 16 (2023): 6074.

Profits in polluting, depleting, plundering industries will plummet, especially doof, fast fashion, advertising, finance, factory farming, travel, shipping, trucking, and cars.

As the next chapter describes, a few celebrities will become Muhammad Alis of sustainability, a few politicians will create the first "Open China" moments of sustainability, and a few CEOs of major polluting industries will be the first Oskar Schindlers of sustainability. Their peers will follow. They will inspire billions. Some industries will go the way of the slave trade. Entrepreneurs and innovators will see opportunity, liberated from coercive markets. No one will regret leaving them behind.

Support for the APPLE PIE Amendment will rise. The amendment will become our WWII-level mobilization and pass with overwhelming support.

Seeing a future with lower energy but more freedom than today, nearly all couples will want at most two children. Global population and consumption will drop to sustainable levels and the economy will likely improve, contrary to popular expectations (see the coda for background on people freely choosing smaller families and the positive economic effects).[69]

Pollution and depletion levels will plummet, though they won't reach zero for a long time due to past activity. Health and biodiversity will rebound. Our power grids will decrease and decentralize. National and personal security will increase.

Remaining fossil fuels will be spared for agriculture, police, hospitals, and redesigning cities to require no more fossil fuels. Elsewhere, those already practicing Sustainable Free Abundant behavior will share what works. Entrepreneurs will develop new systems to deliver necessities.

Cultures will return to Sustainable Free and Abundant culture and away from PAID, which doesn't mean acting or looking like current

---

69   O'Sullivan, Jane, "The social and environmental influences of population growth rate and demographic pressure deserve greater attention in ecological economics," *Ecological Economics*, 172 (2022)

indigenous people. Participation in culture, sports, cooking, government, education, and gardening will skyrocket. Passively watching TV will plummet, as will social media use. We will spend more time with loved ones. We will dance and sing more.

Immigration will decrease as regions once impoverished by imperialism stop seeing their wealth siphoned off. Wealthy regions will realize the best way to help impoverished regions is to stop plundering their resources and addicting them to PAID culture. When rich nations reduce our environmental impact to lower than poor ones, we will be able to lead them. I currently impact less than the average Indian citizen. When all Americans follow suit, assuming poor nations don't reduce their impacts faster, then they will become open to our vision and plans.

Sustainability leadership workshops will spread globally. We will stop clinging to fantasy boomer futures requiring polluting, depleting, and plunder. Accepting that we can't change the past, we will acknowledge that past pollution will cause sea levels and pollution levels to continue rising, and that millions of people will continue to suffer. But we will see that our best course is to do what Teddy Roosevelt suggested: what we can with what we've got where we are, without succumbing to factionalism. We will not evaluate ourselves by things outside our control but by how we perform relative to our potential. We will love doing our best.

Mines and oil wells will close. Imperialism will decrease. Economies will become local. Farmland will become more valuable. We will garden more. Agriculture will return to employing more of the work force and become regenerative. Cultures and communities already Sustainable Free Abundant will be resources for others to learn from. Suburbs and exurbs will return to agricultural land or rewild. Small towns will return.

There will be less electric air-conditioning. People will stop living in unsustainable numbers in deserts and other marginally habitable places. Phoenix and Las Vegas will become small towns, appropriate to what the local environment can support. People will rediscover how our ancestors thrived without them. Birds and bees will return. We will live more in harmony with nature.

We will experience deep emotional journeys facing guilt, shame, tears, helplessness, and having been complicit—overcoming it with action and resolve. We may institute truth and reconciliation committees. The population will become on average older for a decade or two as birth rates decline then stabilize, but since health spans will increase faster than life spans, the fraction providing for those who can't provide for themselves will be able to.

We will enjoy cleaner air, land, water, and food. A basic internet will remain, though with less 8K video than we use today. Hunting will become viable again. Flying will stop being viable unless innovators make it sustainable. People will rarely live flying-distance from their families. Trips across oceans will happen by sailboat. We will enjoy more rewarding travel experiences so won't miss flying. Cultural diversity will return as culture stops becoming homogenized over flying distances.

Innovation, ingenuity, and entrepreneurship will soar, though innovators will find ways to increase freedom and stewardship instead of technologies that pollute, deplete, and plunder. We will participate more actively in civics, arts, culture, sports, and education. Instead of watching them on TV, we'll play sports, tell stories, write, act, paint, and so on. Humanity will become more fit, physically and mentally. We will understand more science, favoring interacting with plants, animals, soil, and so on over abstract theorizing.

As more people freely and joyfully reduce their impacts by 90 percent, those who first hit such levels will reduce another 90 percent, increasing freedom and humanity's time to restore sustainability. No extra suffering or deaths will happen beyond what's baked in from past pollution.

Within a century, through voluntary, non-coercive, joyful choices for smaller families, global population will reach levels at most half of today's, living with qualities of life on par with Europeans today. Most people will again be able to walk to experience solitude and awe in nature when desired. There will be little to no litter or doof, and less of its results like heart disease, diabetes, and obesity. People then will look back at our time not with envy but with horror, the way we look at slavery. They will ask and lament how we could have persisted so long.

Eventually, through democratic, non-coercive means we may choose to pursue unsustainable but possibly desirable long-term pursuits like reaching Mars and beyond.

The Transition

Why act? We know in our hearts but don't want to accept: when we pollute, deplete, and plunder, we hurt people. I've covered the groundwork to help you not only accept, but embrace exiting and ending PAID culture.

If what we do kills innocent people and we don't want to kill innocent people, we have to stop doing what we're doing, even if we like it. If we keep doing it, our inner conflict will tear us up inside. Systemic change begins with personal change. Reversing our death spirals will come from practicing the basics.

I have concluded that even if my behavior could not possibly influence anyone else, even if the changes were deprivation, sacrifice, burden, and chore, I would still have to act to live by my values, the way Franklin, Carter, and Wallenberg did but Jefferson, Madison, and Washington did not. Acting on intrinsic motivation, as the Spodek Method prompts, brings freedom, joy, fun, community, connection, meaning, purpose, and other rewarding emotion.

To recap some basics of stewardship:

- You are powerful, not powerless.
- Participating in PAID culture harms others and yourself more than you realize.
- PAID culture destroys life, liberty, property, and the pursuit of happiness. It corrupts government and markets. It plunders.

- Living unsustainably means you are supporting imperialism and undermining free markets.
- You tell me what you fear losing and I'll tell you what you'll gain more of.
- Exiting PAID culture leads to freedom, fun, joy, community, connection, meaning, purpose, and rewards that have long been human birthrights. You'll be glad you did and wish you had earlier.
- We confuse what life requires with what our lifestyles require.
- Business as usual won't maintain our current situation. It will bring population collapse. Billions could die within our lifetimes. Being rich offers negligible advantage.
- You aren't alone. Others will join. Even if no one else acts, you'll be glad you did.
- Unsustainability led to racism, scientific racism, pollutionism, and scientific pollutionism.
- Living sustainably doesn't mean reverting to the Stone Age, going backward, or a dystopia.
- Cultures often called "Stone Age" aren't. Many are sustainable, free, and abundant. They see PAID culture's medicine, planes, and so on and resist joining out of awareness.
- In head-to-head competition, people in Sustainable Free Abundant cultures choose theirs over PAID culture.
- They see PAID culture as vicious, cruel, overworked, lacking community and don't want to sacrifice their freedom, equality, community, and mutual support.
- People who haven't tried living sustainably don't know what they're talking about regarding moving toward sustainability, including nearly all prominent voices.
- PAID culture is not your fault, but only we can take responsibility to exit it.
- It was reasonable for those who created PAID culture to believe polluting and depleting wouldn't cause the harm it did.

- The transition will start hard, then will become easy and feel natural.
- We don't need to know a full solution in order to start. We know enough already.
- Ramping up solar and wind before ramping down less sustainable energy sources like fossil fuels leads to using both types of sources.
- The worst way to develop alternatives is to keep using PAID things. If you believe we can make air travel sustainable, the worst way to achieve it is to keep flying unsustainably.
- Those participating in PAID culture the most depend on it most so will have the hardest time exiting.
- Proposals from Silicon Valley, Washington DC, and academia tend to augment PAID culture.
- If you make a polluting culture more efficient, you may lower pollution in some areas but will pollute more efficiently, increasing pollution overall.
- Making our culture sustainable, free, and abundant will lead to more innovation, not less.
- The challenges we face and solutions we need are more social and emotional than technological or legal.

In the long run, the Spodek Method, which helps people start from inaction, will be overtaken by community in influencing more people as more people live joyfully sustainably. As sustainability forms community, others won't need to start people from inactivity on sustainability.

Existing communities will embrace and promote sustainability (not just talk about it), particularly in spirituality and faith. Podcast guests who are leaders in Christianity, Judaism, Buddhism, and Islam share how each faith promotes sustainability and alleviating suffering. Those who have lived among Sustainable Free Abundant cultures tell me sustainability runs deep within those cultures too.

CHAPTER 20

YOU, HERE, NOW: WHAT TO DO

Systemic change begins with personal change. You can't lead others to live by values you live the opposite of. To change a culture you must first change yourself.

Even with all of the above reasons, you have to start with intrinsic motivation or you'll always push against resistance, sabotage your efforts, and won't enjoy the process.

I recommend the following process:

1. Start learning the Spodek Method. It will shift your motivation from extrinsic to intrinsic. You can download the free how-to workbook at SpodekMethod.com and begin practicing today.

2. Follow your mindset shift with continual improvement. Keep finding more to act on intrinsically. You ought to be able to lower your impact 90 percent in under three years.

3. Practice the Spodek Method with others. You will see patterns in rationalizations and justifications that reveal your own internal contradictions and corruptions.

4. Join or start a mutual-support group. Acting with others makes you feel like you're swimming downstream.

5. Find role models more advanced and become a role model for those following.

6. Create community.

After you find intrinsic motivation, you'll learn everything you need by simply starting. You don't have to wait or figure out details. Analyzing and planning only delay starting. For the top-down part of the strategy, if you know anyone renowned and interested in helping create a legacy, please contact me directly.

Longer Term

Everyone's experience is unique, but you can expect that you will

- Find each step of living sustainably becoming easier and more natural.
- See a path to a sustainable world more free, safe, fun, and joyful.
- Experience more mindset shifts. (My second one came after Frances Hesselbein's memorial service, when I found not a quirky personal interest in an abstract environment but universal human emotion—in my case love for others, specifically innocent people I would otherwise have hurt—motivating climbing all those stairs. Another came from seeing the Spodek Method bring, more than anything else, liberation and freedom.)
- Increasingly want to learn more about people suffering from our unsustainability. It will increase your motivation, not guilt.
- Feel more resolve and want to share your experiences.
- See techno-optimist and ecomodernist visions like the cotton gin and highways in cities: innovation whose unintended side effects overwhelm its intended ones.
- Learn to share love without flying to visit in person.
- Become increasingly fit, physically active, and culturally active.
- Reduce doof and addiction, eventually to zero.
- Prefer smaller families, not out of coercion.
- Connect more with community, especially in person. You will participate more in politics, especially local.
- Be glad to see entire industries close, as slave plantations did.

- Enjoy owning less.
- See highways, airports, and parking lots rewild or become agricultural as communities reconnect and reform without highways dividing us.

You may also take on increasingly bigger challenges, like organizing many workshops to serve industries or running for office. As a society, we may create truth and reconciliation committees to help overcome guilt and shame.

CHAPTER 21

BEYOND YOU AND ME: THE MOVEMENT

At the time I'm writing, few people are living joyfully sustainably. Success in sustainability will mean billions of people living joyfully sustainably. Let's see why crossing that chasm is achievable. We already know each person who joins will feel liberated and look forward to leading others.

Bottom-Up Precedents

First, let's look at some movements that grew from nothing: Alcoholics Anonymous (AA), Weight Watchers (WW), and Crossfit. They are different than sustainability, but each is challenging but liberating for each new person. Each movement grew through a decentralized process.

When AA began in 1935, many Americans saw alcoholism as a moral failing. Treatment came from religious groups and medical professionals promoting painful interventions and confinement in asylums. A few people developed what became a twelve-step program. It has its shortcomings and critics, but works for many and has expanded to other addictions. It grew through a decentralized approach that enabled participants to help others and to form new groups. By 1950, one hundred thousand people had participated; by 1976, that number grew to one million.

WW began in 1961 when one frustrated woman wanted to lose weight. She invited six friends to share their experiences and practice together. She didn't present facts and instructions. She described overeating as an emotional problem with an emotional solution. Within six months, forty women were attending. By 1968, over five million people had participated.

Crossfit was conceived in 1996. Its first gym opened in 2000. A Crossfit gym didn't require much equipment. They posted workouts online in 2002. Crossfit grew to 15,500 gyms in 162 countries in 2018, with up to five million estimated members.

Alcoholism and overeating are hard to overcome, as is starting a fitness regimen. No program is perfect, but these three have helped hundreds of millions of people. They work through community, mutual support, role models, and other techniques based in social and emotional skills. They distinguish between lecturing facts and helping people solve their problems. They grow by enabling members to lead others and start new groups.

A Sustainability Movement

The Swedish Maja Rosén 🎙 , shared on my podcast how when she stopped flying, most other Swedes saw avoiding flying as abnormal and asked her to explain herself. Within a few years, culture reversed. Enough Swedes stopped flying that avoiding flying became normal and people needed to explain flying.

How Big How Fast?

AA and WW took decades to reach millions of members. Crossfit reached millions in under two decades. Sweden's flying culture flipped in a few years. I could see an intrinsic motivation-based movement for sustainability and stewardship grow yet bigger and faster. Unlike those movements, a life-and-death problem facing every person alive motivates sustainability. The sustainability movement is also starting with hundreds of millions of people marching globally, aching for leadership.

Top-Down Precedents

Imagine if you could look around and see people living joyfully and freely for living sustainably. Imagine they shared their reasons and you found them genuine and authentic, not just greenwashing, but acting after internal searching and external research into what works. Today we're more likely to read how people fly private jets to climate conferences, sports teams fly one-hundred-mile ten-minute flights, and celebrities fly seventeen-minute flights.[70, 71] Elon Musk's private jets took 441 flights in 2023, including from one airport in LA to another in LA.[72]

People crave role models in sustainability. Many well-known people want to create legacies. This match of supply and demand creates an incentive for those well-known people to create those legacies. It will work only if they act genuinely, authentically, and with integrity. So far, I know of nearly none who have. I see people of renown holding back from acting out of fear looking like they're greenwashing or losing votes or sales.

The Spodek Method done publicly achieves what they are looking for. By revealing intrinsic motivation, it shows genuineness and authenticity so when, say, Leonardo DiCaprio practices it, his fans and peers will want to follow him. He doesn't have to be perfect or solve every problem, just genuinely and authentically show he's doing his best. It shares joy.

Consider what would happen if a few people of renown acted. I'll share a vision I see for such people coming from the influential worlds of celebrity (actors, athletes, musicians, etc.), politics, and business.

Once people see an effective way to create a legacy of authentic, genuine action, there will be a race to act first. Those who go first will create legacies. The rest will have to follow.

---

70 John Stanton, David Lockwood, and Katie Gornall, "Premier League: Should Clubs Stop Flying to Domestic Matches for Environmental Reasons?" *BBC*, November 12, 2021.

71 Oliver Milman, "A 17-Minute Flight? The Super-Rich Who Have 'Absolute Disregard for the Planet'," *The Guardian*, July 21, 2022.

72 Grace Kay and Taylor Rains, "Elon Musk's Private Jets Took 441 Flights This Year," *Business Insider*, December 17, 2023.

Celebrity: Who Will Be the Muhammad Ali of the Environment?

I've asked hundreds of people if they know of anyone with a prominent voice in sustainability trying to live sustainably themselves. A few suggest Greta Thunberg. Some will offer someone they know living in the woods, but not with prominent voices or in ways they want to emulate. Plenty of celebrities talk about sustainability. Few live it. We have many people who talk like Thomas Jefferson who could use more action like Harriet Tubman's and integrity like Robert Carter's.

To their credit, some celebrities mean well. Some try token efforts like buying offsets, but end up inauthentic, not genuine or effective. They're stepping on the gas, thinking it's the brake, wanting congratulations, leading their fans to follow with ineffective, superficial acts.

Recall Ali's courageously and selflessly acting on his conscience as I described at the end of chapter 15. I believe it transformed him from merely a boxing champion to one the world's great statesmen. Athletes who take stands today nearly all follow his legacy. There is a legacy like his to be had in sustainability, maybe greater. One of the few American candidates I know of is Ed Begley Jr., an accomplished actor who takes the bus to the Oscars with his daughter. People at the level of Taylor Swift and Lionel Messi could reach Ali's level. They only have to act authentically, genuinely, and effectively on an issue the world already craves for people like them to act on.

I sympathize with them. I doubt any of their handlers are helping them understand our environmental situation, not that celebrity handlers would be prepared to know any better, let alone how to act effectively. It's easier for celebrities to blend in with peers waiting for someone to go first and face the first criticism. Ali did the opposite.

The Spodek Method would lead them to that authentic, genuine, effective change. It enables them to share their journey, so fans hear their vulnerabilities, flaws, and intrinsic motivation. Then they support instead of criticize. Fans have flaws. They want to hear other flawed people succeed, overcoming failure, so they can try too. The Spodek Method doesn't create a Disney happy ending where a simple change lets you live happily ever after. It enables real people to change effectively.

The celebrity who acts first will change history. Others will follow. Those who act with genuine, authentic, effective, bold initiative will create a legacy in an area of global demand that, if humanity survives, will endure centuries, maybe millennia. Their fame today will transcend that of their peers. If you are a celebrity and want to act authentically, genuinely, and effectively, or know one who does, contact me. Let's put you in the history books.

Politics: Who Will Create a "Nixon Opened China" Moment?

We talk of politicians as leaders, but they aren't leading in sustainability or stewardship, so there is space for a politician to become a sustainability leader. They wouldn't act like Ali, though, so I propose Richard Nixon's 1972 visit to China as a more relevant target to replicate.

At the peak of the Cold War, it was "one of the most important turning points in twentieth century history—perhaps the most important in the post-World War II era," said Scott Kennedy, senior adviser in Chinese business and economics at the Center for Strategic and International Studies. "The US and China were able to overcome their antipathy to reach this détente only because of their common foe—the Soviet Union. Short of that, there would not have been this détente."

Then-Senate Democratic Leader Mike Mansfield clarified that "only a Republican, perhaps only a Nixon, could have made this break and gotten away with it." Nixon going to a communist stronghold enabled anybody less anti-communist—that is, nearly anyone—to interact with China too. Several political conservatives and libertarians have moved toward sustainability, including former South Carolina Congressman Bob Inglis 🎙 and former Cato Institute spokesman Jerry Taylor. They are prominent, though not presidents of the US.

Soon a prominent opponent of sustainability will create a "Nixon opened China" moment. If that person is a republican or libertarian, he or she may leapfrog all democrats.

Business: Who Will Be the Oskar Schindler or Robert Carter III of the Environment?

People look to business leadership from companies that present themselves as more sustainable, but the executives with the greatest potential influence are those heading the most polluting companies. They can lead their companies, executives at peer organizations, and the public. So far, they have acted the least (while marketing themselves the most).

I sympathize with them. They fear losing their jobs. As with celebrities, their helpers likely aren't helping them understand our environmental situation, not that their staffs would know better, let alone how to act effectively. They're likely distracted and misguided by people like Elon Musk and Bill Gates, who get attention for stepping on the gas, thinking it's the brake, wanting congratulations. It's easier for them to wait for someone else to go first and follow. Businessmen Oskar Schindler and Ray Anderson did the opposite.

They too can create legacies like Ali's. They also need only act authentically, genuinely, and effectively on an issue the world is craving for people like them to act on. The Spodek Method would work for them too.

It's tempting to think CEOs have only two options to act more sustainably: to shrink the company and its profits—in which case the board or shareholders will fire them—or to quit, in both cases leaving the company more polluting and depleting than before. But successful leaders don't blindly act in the first way that pops into their heads when they think of projects. The successful, effective executives I've worked with reflect, consider their resources, obstacles, options, allies, and so on. They create strategies and plans that may involve dozens of colleagues inside and outside their firms over months and years.

Carter, Schindler, and Anderson didn't just act to make quarterly numbers. They made history because they didn't act first as businessmen and second as human beings or leaders. They acted first with deeper values than mere business interests. They saw that they were acting in coercive markets, not free ones.

CEOs and politicians today could see themselves how we wish more Germans in the 1930s and slaveholders in the US did: as humans first and CEOs second.

What Would a CEO or Politician Actually Do?

Here is a thought experiment exploring what an influential person of renown might do. How such a situation might play out would depend on innumerable details unique to each case, such as the leader's situation, allies, constraints, and so on. Instead of trying to research any actual situation, I'll lay out my assumptions for a hypothetical case.

Let's say the CEO of a major polluting company, a senator, or a president took this book to heart and contacted me. Let's also say that by then a celebrity or two had done so first, as had a large number of non-celebrities so this person saw people actually changing their behavior, not just talking. What would we do?

Let's say it was the CEO of one of the super-major fossil fuel companies, like Exxon, Shell, or BP. Many people expect he (they're currently all men) would be fired if he suggested a change that would reduce profits, but successful executives who plan major actions don't just act simplemindedly. They consider their goals, resources, allies, obstacles, and so on, to create long-term missions and strategies. I'd speak with him as confidentially as he wanted and ask him what brought him to me: his needs, interests, goals, and such. I wouldn't 4C-bludgeon him with science or what I would do. I'd learn his motivations, constraints, relationships, and so on.

Let's say he shared he wanted to find ways to lead in sustainability despite big constraints. Besides connecting and learning what I could, as with any client, I'd do the Spodek Method to start his mindset shift and process of continual improvement in sustainability leadership based on his intrinsic motivations. If no intrinsic motivations arose, I wouldn't have anything to work with, so couldn't continue. I won't try to impose my values or create motivations where none exist. People's greatest motivations tend to be their greatest vulnerabilities, so he may need support to feel comfortable sharing.

Let's say he does the Spodek Method and experiences a connection to deep intrinsic motivations he wants to act more on. Besides exploring the experience with him, I'd likely do the Spodek Method a few more times since people advance with each iteration. I'd have him practice it with others to see and feel what doing it brings to others. If he concludes he wants to leave the field as fast as possible and quit, I could only support him in his interests, though I'd explore them.

Let's say he concluded something extreme, like humanity needs to stop using fossil fuels sooner than he ever thought and he wanted to act personally. Let's say he felt his own company should reduce extraction, but he also doesn't see the point because of objections: if his company extracts less competitors will extract more, his board and shareholders would fire and sue him, and so on. But let's say he also believes someone has to act first, his company could be it, and he could create a legacy that would last centuries.

The challenge will not likely be technological. It will likely be leadership issues, meaning people, competing interests, politics, culture, vision, and so on. I'd work with him to understand his situation, which he knows better than I do. What does he believe he can achieve for his company? What are his constraints, relationships, personal interests, and views of the stakeholders including board members, shareholders, employees, suppliers, customers, unions, governments, competitors, and media? What resources, objections, and alternatives does he see? Only with a calm, clear view of the situation can we create a strategy. We might see no way forward. More likely, we'll see only a hint of one and have to work at it.

For him just to talk publicly about extracting less would explode in the media like a bombshell. The media is more motivated to foster controversy than understanding or effective sustainability work, so would likely make a big, global story out of it. He would then lose control of the conversation. He would therefore have to avoid talking to anyone who might leak the story to the media.

Reiterating that I'm pursuing a thought experiment, I'll explore with broad strokes what we might explore next. We might conclude he could only talk to trusted board members he has deep relationships with and trusts that they would keep this internal talk confidential. We'd likely presume that coercion wouldn't work, so no bludgeoning. We might train him to where he could do the Spodek Method with them, to where he could meet with them one-on-one and lead them to be open to his ideas.

Let's say after months or longer of his leading this internal team, a majority of the board agrees it's in the interests of humanity, including themselves, to follow Robert Carter III and improve the world by extracting less, though also with a plan to lead their competitors to reduce too, as Britain did after banning the slave trade. They would likely then talk to their legal counsel, also with a duty to keep the idea confidential.

Let's say they all tentatively agree, comprising a majority of the board and legal counsel. Now they have a mission. It may still not be possible. Next they could create strategies and a team to implement that mission.

All stakeholders would be affected by any plan, which would suggest involving them all in the process, but any leaks to the media would sink the plan. We might conclude the next step would have to be to go to the media first. This step of circumventing all other stakeholders will surprise them, so a message would have to be near-perfect. If delivered effectively, it could be one of humanity's great speeches. It would depend on an intricate examination of every part of the situation, all relevant relationships, and so on. Though it would benefit from input from all stakeholders, that input would be impossible, so it would have to be created by the board members and lawyers involved. They might take months to create, edit, and practice it. They might still conclude it's impossible and stop. Or they might go for it.

The Speech

Continuing my thought experiment, I'll explore a possible speech. For an example of such a speech given by the CEO of a global corporation, see the 2020 Regeneration Speech by Doug McMillon, CEO of WalMart.[73]

Imagine a world where this movement had taken root to where consumers had already voluntarily chose to reduce flying by 20 or 30 percent with signs of more to come, doof and bottled water down by 80 percent, electric grid use down 20 or 30 percent, container ships and truck shipping down by 30 or 40 percent, and social media down by half; accompanied by increases in library use, sports participation, community theater, home cooking, and gardening. If the results brought joy and freedom, executives at corporations that undermine stewardship might see they have to change.

As you'll see, I expect it will be more effective to say he and his company are vulnerable than to say they're powerful or have all the answers. It might go like this:

> Hello and thank you.
>
> Coal, oil, and natural gas have enabled a way of life for billions of people nobody could have imagined before. With them we have built taller, traveled farther, fed the hungry, and lifted billions out of poverty. Nobody questions these marks of progress. We have always known that these fuels are finite and pollute. Now, humanity must wean itself off them.
>
> There are no easy answers or next steps, but research has shown conclusively that to continue business as usual is no longer acceptable. Billions of lives are at stake.

---

73   2020 Regeneration Speech—Doug McMillon, President and CEO, Walmart.

For decades humanity has known it must act, but we at [company] can't act unilaterally.

Despite being consistently rated a well-run company, with best-in-class employee safety and satisfaction, satisfied repeat customers in all our businesses, generations of exceeding market expectations, and research and development second to none, we must acknowledge our vulnerability. We deliver power, but we are not all-powerful, especially to solve humanity's fuel problems. We cannot do it alone. To do what benefits humanity and life on earth most, we need help.

We cannot think of a better next step after decades of research and thoughtful consideration than to ramp down production from existing sources. Therefore, we will stop all exploration and will dig no new wells or increase capacity. We will remain an energy company for energy we can supply genuinely cleanly. We will transition to becoming what the world needs most: a company transitioning away from polluting and depleting.

We intend to work with all of our stakeholders, as any jobs and businesses will be affected by this action and markets will react dramatically. Most of all, we agree with the science that says that if humanity doesn't ramp down to zero fossil fuel extraction now, the alternative is worse than lost jobs and share price.

We believe our action is consistent with the purpose of the corporation and law. Shares are worthless in a world that can't sustain civilization.

Regarding jobs, pensions, and the savings of working people invested in our company, we have all long known humanity would have to face these challenges. Facing them decades ago would have been easier. Waiting until tomorrow will make it harder, even impossible. We hope you will help us with our people in jobs that depend on extraction. We will help with severances and retraining, but if society agrees that extraction must end for our collective health and safety, we hope to work with governments, entities in sustainability fields, and others to help put the best workers in the world back to work.

While no situations are identical and we want emphatically to avoid false equivalences, the most relevant example we could find of a global entity ending a practice it depended on since its founding was Britain banning the slave trade in 1808. Prior to it, the overwhelming consensus was that if they didn't do it, others would and the empire would collapse. Instead, all the other slave-trading nations joined. Then all banned slavery. The practice exists, sadly, but is nowhere legal. Again, we do not want to create false equivalences, only to point to precedent of competitors joining, not taking advantage of our vulnerability.

The leadership of [company] are not personally benefiting from this action. I am pegging my salary to our lowest paid employee. We have prepared more detailed written plans that we will collaborate with our stakeholders to complete and implement, as well as the media and public. We hope you'll keep in mind that we created our vision, strategy, and plan as a call for help and support to come together for a productive, reasonable conversation and action in this new day for humanity.

Our only benefit is all of our shared benefit in cleaner, healthier, safer air, land, water, and food; and a more secure world. This action has been necessary for too long.

EPILOGUE

"I've thought about these problems for years and years. The Spodek Method is what I needed to finally do something!" Someone Evelyn taught said these words—a guy I've never met. He continued, "People think that sustainability means more expensive. But I'm saving so much money! And I still have more free time. It's mind blowing."

He feels joy and gratitude so he'll soon train others, who will train others, and so on. This movement is spreading without me.

Before the workshop, Evelyn knew about our environmental problems but had given up. During the workshop, she had struggled. She insisted that because she knew less science and hadn't done anything in the field of sustainability before she couldn't catch up to everyone else. She was divorced with three boys in joint custody, meaning no extra time or money. When she practiced the Spodek Method, she couldn't get it to work.

But she kept at it. Then it clicked and she saw sustainability leadership as what she had looked for her entire life, especially since becoming a mom. As a result, she now spends more time with her sons doing more active things. The resulting meaning and purpose, plus mastering the technique, improved her relationship with her family and ex-husband.

Now she's teaching others and they're telling her their equivalents. Soon, some of them will teach their cohorts and hear from the people they lead and inspire, which might be you.

You read Beth's recommendation on page 20. Wouldn't you love to feel inspired to write such a recommendation, especially about something you cared deeply about but had almost given up on? Wouldn't you love someone to write such words about you for having led them? You can do both. Through the Spodek Method how-to workbook, a free download from SpodekMethod.com, you can learn to lead yourself and others. Advanced courses can enable you to lead your own workshops and bring liberation and freedom to people like you. They will thank you for it. They will write recommendations about your leadership as Beth did of mine.

Evelyn kept saying to me, "You said this would happen," each time she found liberation or joy where she expected feelings of loss or deprivation, or someone would thank her for leading them to do what she thought would feel like pressure or burden. She'd continue, "I didn't believe you, but it keeps happening." She still says it. *I* still say it as I find yet more freedoms and ways to appreciate nature and people.

There is no way ten years ago, on the day before I started avoiding packaged food for a week, that I could have predicted doing the experiments that led to writing this book. I would never have predicted that I would:

- Be the first of four billion people (that I know of) to disconnect from the electric grid while living in a city;
- Discover a twelve-thousand-year history leading to our situation;
- Find no bad guys, only procrastination;
- Discover liberation, freedom, joy, and fun;
- Refine a way to bring others these results; and
- See them lead yet others to these results.

One of my most liberating and freeing discoveries was to see that there are no bad guys perpetrating our problems. Sabrina and Kevin didn't intend to start dominance hierarchies. Twelve thousand years ago circumstances happened to make such transformations possible that nobody could have foreseen to create our PAID culture.

There are no good guys asleep at the wheel though, either. We are the only ones who can engage and act to lead ourselves and others to change global PAID culture to restore stewardship. If we only do what we're good at now, then we will all "do our parts," feel good about ourselves, and keep accelerating the problems.

I have heard many reasons not to act on sustainability. The day before my week avoiding packaged food I believed all of them. Midweek, I may have believed them even more.

Since then, people have told me many reasons they don't act. All sounded sincere. All sounded like they had no choice but to keep doing what they were doing. Each reason made the person sound like one of the good guys. No matter how unique their reasons sounded to them, I believed a version of every one of them for myself and heard others say them too.

A main part of any culture is its unquestioned, unchallenged, untested beliefs. No matter how true a person considered their reasons, not one of them had meaningfully questioned, challenged, or tested their beliefs. They were as confident, passionate, and wrong as slaveholders who proclaimed, "Freedom is not possible without slavery." They were as confident, passionate, and wrong for the same reasons: they held status in a dominance hierarchy and knew in their guts that if they didn't maintain their status that "complacency can be fatal."

We fear losing our lifestyles or even our lives if we act enough on sustainability to change culture. So we say, "sustainability is important. I will act on it. I just have to do this one thing first," where that thing is our job, family, or other unassailable value.

In other words, we procrastinate instead of engage. We see engagement in sustainability as distraction from higher values and therefore deprivation and sacrifice. Procrastinating on sustainability feels tantamount to saving our lifestyles or even lives.

My experiments questioned, challenged, and tested these cultural beliefs and found them wrong. What I can share nearly uniquely, as far as I know, is that engaging only looks like distraction, deprivation, or sacrifice. Procrastinating only looks like saving our lifestyles or lives. I know from practical experience including my own and that of world-renowned leaders and regular working people, not just book knowledge or fear-based rationalization and justification.

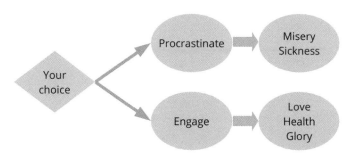

In 2024 and beyond, I know from practical experience that, post-mind-set shift, engaging on sustainability brings intrinsic reward, especially liberation. You will wish you started earlier. When skilled in the Spodek Method, you will want to share and lead others because you know they will feel gratitude for liberating them. Regarding procrastinating for other values, I hope this book has shown that for whatever your values, sustainability is upstream from them. Only with clean air, land, water, and food can we have health, safety, security, family, community, the economy, anti-racism, free markets, and democracy, to name a few.

Start the Spodek Method. You may struggle the first time, but keep at it. Humanity needs your help, your experience, and your connections. Act. Strive until you say in your way what Tubman did:

> If you hear the dogs, keep going. If you see the torches in the woods, keep going. If there's shouting after you, keep going. Don't ever stop. Keep going. If you want a taste of freedom, keep going.

Then:

> When I found I had crossed that line, I looked at my hands to see if I was the same person. There was such a glory over everything; the sun came like gold through the trees, and over the fields, and I felt like I was in Heaven.

CODA: SCIENCE, FREEDOM, AND ART

The Science

"Wait," you might think, "doesn't sustainability start with science? Shouldn't you start the book with science?"

On the contrary, after the mindset shift, the science becomes the wind beneath your wings. Before, it weighs many people down. After, it motivates people to act in ways to liberate themselves, leading them to want to learn more. Before, it makes many feel guilt or shame.

It's like second-hand smoke: once you know it causes cancer in others, you don't need to know the exact mechanisms to know not to impose your smoke on people who can't consent.

I'll share a story to illustrate why not to start your action with science, but rather your science with action. Then I'll share what books and resources I recommend to learn more, followed by how I think about the main scientific points after having practiced and digested them for years—that is, I'll share my view.

Why Act Before Learning Theory

I never expected to become a successful artist, but in the early 2000s, I had two solo gallery shows in New York as well as group shows across the country and Europe. I showed pieces in museums, collectors' homes, Art Basel Miami Beach, VIP rooms of popular clubs in New York and Miami, big public venues, and more. I taught art classes at NYU and Parsons.

It came unexpectedly through the medium I created for Submedia that showed moving pictures to people moving past it. I started by showing it to friends as a technology. Without experience creating art, I didn't know how to make images that people would find beautiful, so I made the displays and partnered with artist friends to make images.

A friend knew the owners of a club and arranged for them to see it. They loved it and we installed pieces in the club's VIP room. A gallery owner saw it and invited me to show in his gallery and things took off.

I started making images myself, but viewers responded less to them than ones by friends. I had never tried to express myself visually so feared showing something I considered beautiful and people laughing at it. Still, I kept at it, sometimes copying my friends' ideas, other times experimenting. Occasionally, people would say one of my images looked good. I'd pursue those directions. Little by little, my skills developed and people responded more positively to my work.

The simpler the images, the more I could create and express something coherent, plus they were easier to work with. I kept cutting out inessentials and reached the basics: I created images using lines, circles, squares, and simple shapes moving in simple motions with basic colors. Working with basic elements led me to discover aspects like composition, rhythm, and texture, though I lacked the vocabulary to name them then. One day I looked at a series of pieces I made from these basics and thought, to my great surprise, "This is art. I created art." Then I started learning theory to see what more I could build with these basic elements.

School had taught me art theory in classes in art history, art appreciation, and some hands-on classes in drawing and painting, but the classes were nearly always required—that is, coerced—and that theory never translated to me creating art. My teachers likely hoped teaching theory would lead me to make art, but in practice, making art led me to want to learn theory, *when intrinsically motivated*. Coerced practice can in principle lead someone to learn theory if it happens to resonate, but that's luck. Starting with intrinsic motivation may not make that desire inevitable, but it plays a big role.

What I found in art applies in sustainability leadership, where science is the theory: teaching science rarely leads people to act, especially when coerced. By contrast, action leads to learning the science, especially when the action is intrinsically motivated. Hence, I believe I can help the most not by teaching you more science before you've done the Spodek Method enough times, but by pointing you to resources for when your appetite arises. I will also share the intuitions I've developed after digesting the science, which I hope will whet your appetite.

My goal is not for people blindly to follow or copy me. Only you can figure out how living joyfully sustainably looks for you. I want to liberate and empower, and I've found in sustainability leadership, that **teaching theory rarely leads to doing; doing leads to learning theory.**

Starting with intrinsic motivation also makes the effort more rewarding. Learning to make art myself made seeing art at museums come alive for me, among many other benefits. Likewise, practicing sustainability made environmental science come alive for me.

By contrast, remember my crying when being made to practice violin growing up? My parents made me take lessons but I don't remember them first finding and evoking intrinsic motivation. Likewise with little league baseball and soccer, so I didn't take to sports until college.

The Need for Environmental Science Is Lower Than You Think

It's tempting to consider the most important thing to learn to help the environment to be environmental science. The science is important to understand the problem and confirm it exists, but not to act. Learning about lung cancer doesn't help many people stop smoking. When I tore my meniscus, learning anatomy didn't help. Physical therapy did. Thus, I've found the most important and relevant science to sustainability leadership is what helps us lead ourselves and others is anthropology and psychology. These sciences tell us that:

1. Human cultures have thrived practicing more freedom and equality than PAID culture
2. PAID culture is just the latest incarnation of a system of tyranny that resulted from material conditions twelve thousand years ago
3. Our PR-firm-on-steroids riders convince us to fear alternatives to PAID culture

I have found understanding these points helps us act by demolishing limiting beliefs. I hope I treated them sufficiently below to make sustainability appealing.

Environmental Science Needs Only to Show We Are Destroying Life, Liberty, and Property and Plundering Today

Regarding policy and changing culture, if you believe that government should protect life, liberty, and property and prevent plundering, the main question environmental science needs to answer is if a behavior violates those values.

To justify the APPLE PIE Amendment, consider pollution in the form of mercury dumped in groundwater or leaded gasoline. Overwhelming evidence shows they destroy life, liberty, and property. They deprive people of freedom. We have made them illegal. Allowing them grants special privileges—"The greatest source of inequality," as Friedman identified.

All environmental science needs to show is if a behavior deprives others of life, liberty, and property without their ability to consent—which I have been calling "polluting"—or depriving future generations of resources, as Locke, Jefferson, and indigenous wisdom described—which I have been calling "depleting." To keep polluting, depleting, and plundering legal will keep leading markets to grow coercion and tyranny instead of liberty and freedom.

I'm not denying that behaviors that polluting, depleting, and plundering allow come from compassion and desire to improve people's lives. Slavery began with those motivations too. Maybe James Watt and other early polluters didn't know their pollution would persist and accumulate for millennia. Maybe they suspected but looked the other way, hoping future generations would solve their problems or were blinded by cognitive biases. However they justified starting the pattern, the result is that polluting and depleting coerce people and deprive them of life, liberty, and property today.

Learning Environmental Science for Sustainability Leadership

Whether you know a lot of environmental science, as I did before starting my experiments, or know little, I recommend learning and practicing the Spodek Method first. The world is full of all the relevant science you could want. I can't write anything you can't find elsewhere. If people haven't found it or it hasn't motivated them, the problem likely isn't a lack of science, but using analyzing and planning to procrastinate, as I did for decades before acting.

It's tempting to say we're already acting on sustainability and for intrinsic motivations; we may not be acting as much as we could, but we have the right balance for ourselves. When someone does the Spodek Method with us, we find it uncovers more motivations significantly more intrinsic and that our old motivations were more coerced than we realized. Then the next few times we lead and are led through the Spodek Method, we find increasingly powerful motivations.

It's also tempting to say, "I should learn more science before starting." You know enough for the Spodek Method. Beware using analysis and planning as I did for those six months between thinking of avoiding packaged food and starting, when just acting led to resolving all my concerns and anxieties. Again, it's very tempting to feel you haven't learned enough to act. More likely, you haven't acted enough to learn.

I recommend reading this coda once, but holding off on following up with the books, videos, and resources until you've done enough reps—that is, enough iterations of the Spodek Method.

How many? Different people move at different paces, but five top signs that will show you you're acting on powerful, deep intrinsic motivations:

- **Get to, not have to**: You feel you *get to* act more sustainably not *have to*.
- **Guilt and shame guide, not hide**: Feelings of guilt and shame guide you, not make you feel like you have to hide them.

- **Joy, not obligation**: You want to share your results and help others follow because you expect them to feel joy and gratitude, not obligation.
- **No more excuses**: You see old reasons not to act as rationalization and justification.
- **You tell me what you fear losing, I'll tell you what you'll get more of**: You see that what you thought you'd lose, you get more of.

Some people get there in three or four reps. Some take twenty. Some experience something different than these five signs, but these five are the most common.

Freedom and Science

What does science have to do with freedom? What do either have to do with sustainability?

A story two boys tell in the documentary *The Hadza: The Last of the First* illustrates this connection.[74] They're Hadza, in Tanzania, about eight and ten years old. They look too shy to look at the interviewer or camera, playing with something in their hands to distract themselves. They tell of being brought by jeep from their home camp to a western-style school. The documentary had already showed that Hadza boys are given bows and arrows around four years old, then they begin hunting small game.

The older boy recounts, "After being taught how to write, we were left to write on our own. Then the teacher came and said, 'That's not how it's done. You're doing it wrong!' We were beaten. We were not used to being beaten at home . . . We slept there one night, woke up very early, and escaped in the early morning . . . It took us two days to walk home. On the way, we dug a watering hole for ourselves. We ate baobab fruit that we gathered, and spent the night at another Hadza camp."

---

74 *The Hadza: The Last of the First*, directed by Bill Benenson 🎙 (Benenson Productions, 2014).

We were not used to being beaten at home.

Could you walk home two days from some place you've never been? People I know would start a two-day trek by visiting a sporting goods store to buy hundreds of dollars of gear, consulting GPS, preparing food, contacting emergency contacts, and maybe getting a medical checkup.

They just started walking.

How? They grew up not attending schools or learning literacy, but learning nature. Science is the study of nature, so they learned science— practical knowledge, not theory or lab research.

The interviewer asked, "How did you find your way home?"

The older boy responded, "We just walked home. It was easy. We are Hadza."

As he says, "We are Hadza," he looks up as if to imply, "We aren't helpless like you"—not as a put-down, as I see it, but incredulous that someone wouldn't know how to walk from one place to another. If science is learning about nature, these boys practiced more science than any scientist I've met, including eight Nobel laureates.

Hadza aren't unique. Jared Diamond also opened his book *Guns, Germs, and Steel* stating: "Modern 'Stone Age' peoples are on average probably more intelligent, not less intelligent, than industrialized peoples," elaborating on how indigenous people—whom he studied for thirty-three years by then, earning a National Medal of Science—can solve practical problems of the natural world better than he can.

Learning Nature—That Is, Science— Gave Them Freedom

Learning about nature—that is, *science*—gave them freedom. I don't mean lab-coat peer-reviewed science. I mean knowing about nature enough to live by. I'll highlight some of the freedom science afforded these kids.

First, it gave them the freedom to walk away. Without that freedom, they would have been forced to accept being beaten by teachers who are, effectively, PAID-culture colonizers.

Second, ancestors for tens of thousands of years enabled them that freedom by stewarding nature. They practice the stewardship that Locke and his peers intended.

Third, the neighboring groups that hosted them overnight gave them freedom to move about freely by extending mutual aid. The host group could because they live in abundance by living within nature's limits.

In contrast, this map from the documentary shows Hadza territory shrinking from 1940 to 2011. Other cultures are coercing them out of their territory, particularly PAID culture, plundering and depriving them of life, liberty, and property without their consent.

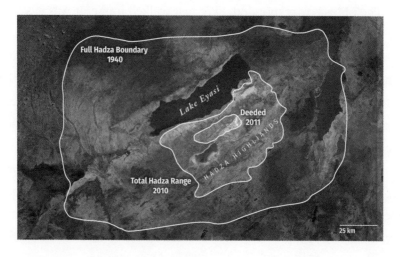

You've seen comparable maps of imperial expansion seizing lands of indigenous peoples. European empires carved up Africa in the "Scramble for Africa." America's colonists and settlers steamrolled North America in "Manifest Destiny." Empires steamrolled from the Fertile Crescent across Europe and Africa, from the Indus Valley across the Indian subcontinent, and from the Yellow River across East Asia.

This last bit of territory in Tanzania marked "Deeded 2011" is one of the last bastions of freedom from PAID culture where someone can enjoy the freedom to walk away if they want, along with a few scattered territories including elsewhere in Africa, near the Amazon River, and on Pacific Islands. By contrast, if I walk two days in any direction from home, I won't even escape the sound of helicopters.

By contrast, I've seen scientists cry at the losses to people and wildlife their research finds. They cry because they see nothing they can do. They lack practical knowledge, skills, and experiences to guide them. Outside their narrow focuses, they are helpless. They pass that condition to PAID culture. Nearly all approaches to our environmental problems promote solutions based on technology, markets, and laws that result in stepping on the gas, thinking it's the brake, wanting congratulations. They mean well, but they don't see how to escape PAID culture. Despite their intent, their lack of hands-on experience practicing sustainability keeps us within PAID culture, accelerating its results.

Freedom Requires Stewardship. Stewardship Requires Science.

Allowing polluting, depleting, and plundering leads markets to grow coercion, tyranny, and inequality. Such markets divert resources from people who solve problems best to people who coerce the most, as slavery did.

Many people hope and claim that we can stop polluting and depleting in proposals they call "substitution," "decoupling," "dematerialization," "circular economies," "green growth," and "clean," "green," or "renewable" energy. They claim that because technologies are becoming more efficient, one day they'll stop destroying life, liberty, and property and plundering. Even the most ardent supporters of these proposals acknowledge they deprive people of life, liberty, and property without consent and plunder *today*.

If you believe container ships or planes may one day operate sustainably then making them legal *after* that transition, if possible, would grow liberty and freedom. Allowing them to operate today grows tyranny. Moreover, allowing them *delays* this transition, if possible, by satisfying today's demand with polluting container ships and planes. Today, airlines confuse consumers with false and deceptive claims of sustainability, delaying innovators from creating sustainable solutions, if possible.[75, 76] If sustainable solutions are not possible, then, like slavery, the faster we disallow them, the faster we end their coercion, tyranny, and inequality. In the meantime, coercers and tyrants wrongly point to their wealth as evidence that they solved problems best. They coerced, plundered, and deprived people of freedom most.

As with slavery, the solution to the coercion and tyranny of polluting, depleting, and plundering is not to slow them by taxing them, which would grow government and motivate more polluting, depleting, and plundering. Nor would developing technology to solve the problems more efficiently, as efficiency would accelerate and grow them. Nor would solving piecemeal the problems coercion and tyranny create. Technology, markets, and taxes don't stop coercion and tyranny.

Only disallowing the tyranny and coercion stop them. In coercive markets, technology, markets, and taxes accelerate them, however well-intentioned. Passing the APPLE PIE Amendment enables people to trade freely, which will grow freedom and equality, not tyranny. Immediately ending pollution, depletion, and plunder would be no more extreme than

---

75 Ajit Niranjan, "Dutch Airline KLM Misled Customers With Vague Green Claims, Court Rules," *The Guardian*, March 20, 2024.

76 Airlines Facing "Greenwashing" Litigation on Three Continents," Jones Day, August 2023.

the Thirteenth Amendment's immediately ending slavery. Loopholes would be exploited.

You may say, "But the APPLE PIE Amendment prevents the market from growing." It prevents a *coercive, non-free, plundering market from growing coercion and plunder*, which deprives people of freedom, and creates more inequality. The Thirteenth Amendment also prevented a coercive, non-free, plundering market from growing coercion and plunder. We're glad it did.

You may say, "But the APPLE PIE Amendment prevents me from doing things I want." Yes, like the rest of the Constitution, it prevents you from doing things you want that plunder or deprive others of life, liberty, and property. It doesn't prevent you from doing things that don't plunder or deprive others of their life, liberty, and property.

How More Science Can Help

I don't mean to imply I see science playing a smaller role in our sustainable future. One of the biggest roles I see is restoring most people's hands-on, how-to knowledge of nature, like the two Hadza boys. Another role, *after* the APPLE PIE Amendment, or at least a cultural shift making it inevitable, is to help find solutions. Only after that cultural shift would science and innovation resulting from it accelerate that culture of stewardship instead of stepping on the gas of today's culture.

Again, since laws permit polluting, depleting, and plunder, any delay passing the APPLE PIE Amendment delays freedom and free trade, augmenting tyranny, increasing the likelihood of today's equivalent of the Civil War, which might be global wars over resources.

Science Resources

Why I Recommend the Resources I Do

Want to know how I know when I read Jonathan Haidt's *The Righteous Mind*? I know it was 2016 because of how I acted on one of its main points—the value of talking to people with different views. In particular, when Donald Trump won the election, I realized living in Greenwich Village meant nearly no one around me voted for the incoming president. Few who did would say so out loud since, while my neighborhood promotes inclusiveness and tolerance, few showed much tolerance for Trump supporters.

I wanted to talk to a Trump supporter. I had a column at *Inc.com* at the time and wrote a piece, "If You Voted For Trump, Let's Meet."[77] I requested people to contact me so I could expand my horizons. My editor there told me *Inc.* doesn't post articles like it, but the policy was not to remove pieces once posted.

Several people contacted me, which led to conversations. I found them wonderful, educational, and different than any I'd had with neighbors. So began my practice of seeking out people I disagree with to burst information bubbles. I'm not the most open-minded person in the world, but I've tried to seek out books and resources on sustainability from diverse views. That doesn't mean I find them all compelling, so I'll list them by what I find uncompelling and what I find compelling.

Most of my life, especially since learning physics, I saw culture advancing through stages marked by energy coming from wood, then coal, then oil, then gas, then fission. I believed each stage provided more energy with less pollution than before. Fusion provides the most power in the universe, including the sun's, so I saw making fusion practical as solving our problems, at least for thousands of years, maybe millions. As solar and wind prices dropped and availability increased, I saw them as renewable with little impact on the environment. I saw people as solving problems, so I felt that a larger population meant more people solving more problems.

---

77 Joshua Spodek, "If You Voted for Trump, Let's Meet," *Inc.*, November 30, 2016.

Books I Find Incomplete or Uncompelling

Thomas Malthus, *An Essay on the Principle of Population.* Studies in population before we learned to analyze populations in terms of systems that included feedback and delays were like biology before evolution, or physics before calculus. They found interesting ideas but lacked the sophistication necessary to cohere into a meaningful whole. Like transportation engineering before gears and brakes, people in those fields created the equivalent of a skateboard. It could transport you, but a bicycle with gears and brakes does better. Malthus wasn't right or wrong so much as exploring new territory. Though more advanced, I include *The Population Bomb,* **by Paul Ehrlich** in this category.

Literature of recent decades, including the ecomodernist and techno-optimist ones I've found, says that more people means more solutions, but as far as I can tell, their analysis lacks a systems perspective including feedback and delays. Their level of analysis lacks predictive power to discriminate between possible future outcomes and ability to explain past counterexamples (like cultures that grew with free trade but collapsed). They can't disprove competing hypotheses.

Their claims against outcomes predicted by complex systems analysis, as far as I've found, are to mischaracterize works like *Limits to Growth,* which I describe in the next section. *Limits to Growth* presents several simulations. Its critics nearly always call simulations predictions, focusing on simulations showing results they don't like or that didn't happen. They ignore simulations that show thriving.

The point of simulations is to explore parameter space to understand the effects of choices on a system. Only by exploring parameter space, including those presenting undesirable outcomes, can we understand the behavior of a system and the consequences of choices and options available. It would be dishonest for researchers not to show a representative range of outcomes. It is dishonest for critics to misrepresent this comprehensive exploration, or ignorant of them if they didn't understand it. As it happens, fifty years of data are consistent with some *Limits to*

Growth simulations, casting doubts on the "evidence" of critics.[78]

Some literature that looks at only one manifestation of PAID culture, such as only climate change or only biodiversity loss, misses the effects of systems and culture. Talking about only climate tends to disregard other environmental problems so that proposed solutions augment them. The problems result from our behavior, which results from our culture. Treating climate only will increase our total problems, with rare exception.

Among books that focus on climate, few distinguish between carbon cycles within the biosphere and reintroducing new carbon into the biosphere from fossil fuels that have been outside the biosphere for tens of millions of years or more.[79] If they confuse these cycles, they'll confuse carbon *emissions* with carbon *newly introduced* for the first time in human existence to the biosphere. I'll return to this topic in the section below on my intuitions.

Julian Simon expressed the idea that more people solve more problems:

> Adding more people causes problems. But people are also the means to solve these problems. . . . The ultimate resource is people—especially skilled, spirited, and hopeful young people endowed with liberty—who will exert their wills and imaginations for their own benefits, and so inevitably they will benefit the rest of us as well.[80]

This view influenced many writers, leading many to see suggestions that earth could be overpopulated as anti-human. But this view recalls the nineteenth century belief "rain follows the plow" that farming would bring rain to North American and Australian deserts. In the words of Charles Dana Wilber:

---

78 Gaya Herrington, "Update to Limits to Growth: Comparing the World3 Model with Empirical Data," *Journal of Industrial Ecology* (2020).

79 Joshua Spodek, "Know the 2 Carbon Cycles and Don't Confuse Them," July 25, 2021.

80 Julian L. Simon, *The State of Humanity* (1996).

God speed the plow . . . By this wonderful provision, which is only man's mastery over nature, the clouds are dispensing copious rains . . . [the plow] is the instrument which separates civilization from savagery; and converts a desert into a farm or garden . . . . To be more concise, Rain follows the plow.[81]

Several correlations between weather patterns as settlers moved west seemed to support Wilber's claim. His explanation seems to precede Simon's:

> In this miracle of progress, the plow was the unerring prophet, the procuring cause, not by any magic or enchantment, not by incantations or offerings, but instead by the sweat of his face toiling with his hands, man can persuade the heavens to yield their treasures of dew and rain upon the land he has chosen for his dwelling . . . The raindrop never fails to fall and answer to the imploring power or prayer of labor.

Only one problem: "The theory is considered junk science, on par with phrenology, séances, and fad diets."[82, 83] It's tempting to confuse correlation with causation and past performance with prediction of future results, but they aren't. Still, they lead to a lot of writing with tempting beliefs.

Literature in this category includes *Superabundance* by **Marian L. Tupy** and **Gale L. Pooley**, *The Ultimate Resource 2* by **Julian Simon**, *The Moral Case for Fossil Fuels* and *Fossil Future* by **Alex Epstein**, *How to Avoid a Climate Disaster*, by **Bill Gates**, *False Alarm* **by Bjorn Lomborg**, *Apocalypse Never* by **Michael Shellenberger**, and *Unsettled* by **Steven Koonin**.

These books generally value freedom, which depends on markets being free, but don't seem to acknowledge that markets that pollute, deplete, and plunder are coercive, not free. I don't see them addressing

---

81 Charles Dana Wilber, *The Great Valleys and Prairies of Nebraska and the Northwest* (1881).

82 James A. Garfield and 'Rain Follows the Plow'," National Parks Service.

83 Henry Nash Smith, Rain Follows the Plow: The Notion of Increased Rainfall for the Great Plains, 1844 –1880, *Huntington Library Quarterly* 10, no. 2 (February 1947): 169–193.

that pollution, depletion, and plunder *happening today* deprive people of life, liberty, and property *now*.

"Rain follows the plow" led to people using technology to try to override nature. It looked like it worked for a while, but nature never ends, so required ever more investment. The long-term results? "The American West quietly became the first and most durable example of the modern welfare state," including "the country's foremost examples of socialism for the rich," as industries that were "grateful recipients of a federal program . . . began to view the benefits as an entitlement, even a birthright. They came to believe that they should receive federal funds and resources in perpetuity" and "the cycle of boom and bust in the West is directly tied to water availability."[84] Do we want to risk human population and a global economy following a boom and bust cycle outside our control?

Even if the people suffering are thousands of miles away, a coercive market undermines our culture here and now, even if we enjoy material prosperity and longevity. Friedrich Hayek's assessment of the slide into totalitarianism from not protecting freedom and liberty in *Road to Serfdom* applies here, but these books seem willing to compromise on freedom and disregard coercion.

If this literature and its authors genuinely believe free markets, free trade, and human ingenuity solve problems, then *first* promote removing coercion and deprivation of life, liberty, and property, *then* let people trade. The other way doesn't work. Coercive markets don't create more freedom. That's why the Thirteenth Amendment unifies us today and why the APPLE PIE Amendment will too. If they believe substitution, decoupling, dematerialization, and so on are possible, the APPLE PIE Amendment will empower innovators to make these processes happen faster—or at all. If an innovation makes something no longer pollute, deplete, or plunder, *then* allow it, not before, when it will deprive people of life, liberty, property, and therefore freedom. If an activity cannot be

---

84 Marc Reisner, *Cadillac Desert: The American West and Its Disappearing Water* (1986).

made to avoid polluting, depleting, or plundering best to make it go the way of slavery. Otherwise, we'll move toward today's equivalent of the Civil War.

If these authors don't believe it's possible to substitute, decouple, dematerialize, and so on but they keep promoting it, what else can we conclude but that they are lying or at least ignorant or deceiving? *First* decouple, substitute, and dematerialize, *then* allow the no-longer polluting, depleting, plundering activity.

It's tempting to say we should allow companies to keep operating as they are so they can solve their problems. If they believe it's possible to make polluting, depleting activity sustainable, but only by promoting more polluting and depleting, I suggest they clearly state, "We believe the ends justify the means. We believe that coercive markets can achieve what free markets can't so we support coercion over freedom."

What if we keep pursuing growth in the belief that more people solve more problems, as rain was to follow the plow? As science writer Matt Simon wrote, "It's one of America's least known and most brutal ironies that 50 years after the pseudoscientific frenzy to plow the Great Plains into submission, excessive tilling led to the Dust Bowl."[85] Do we want to repeat that experiment with humans?

Books I Find Compelling

To understand the science that matters, I recommend starting with the following:

First, **Tom Murphy's** 🎙 **2021 textbook** *Energy and Human Ambitions on a Finite Planet: Assessing and Adapting to Planetary Limits* (a free download), which I consider the science book of the decade. It covers the relevant science from first principles, including economics, population, space, comparing various energy sources, transportation, and adaptation.

---

85   Matt Simon, "Fantastically Wrong: American Greed and the Harebrained Theory of 'Rain Follows the Plow'," *WIRED*, June 25, 2014.

Tom's background is relevant. He has a PhD in physics from Caltech and practiced as an astrophysicist at University of California San Diego until retiring in 2023. He led a team that designed and built lasers to bounce off reflectors left by astronauts on the moon.

He wrote on the book's origins: "I myself first approached this subject—when assigned to teach a general-education course on energy and the environment—with great enthusiasm, intending to sort out to my own satisfaction how I thought our gleaming future would migrate to renewable energy." He found no textbook covering the subject from first principles so he started a blog called *Do the Math* on it. He began experimenting living more sustainably and documenting his results, so he filled at least two of three zones of the Venn diagram of sustainability leadership. I don't know how much he was trying to lead. His research surprised him about how dire our situation is and how counterproductive nearly all our proposed solutions are. The math matters.

His book presents the math to understand our problems and proposed solutions, at the level of a non-science undergraduate. He cites his sources. His goal is for you to understand.

Among his results:

- Decoupling, substitution, and efficiency gains don't solve our problems and can augment them.
- Space exploration won't solve our problems and isn't useful for more than tourism.
- "Fossil fuel use must fall back to essentially zero in a relatively short time (within a century or two)."
- "Because of the downsides of fossil fuels and the inferiority of the substitutes on a number of fronts, it is unclear how we might patch together an energy portfolio out of the alternatives that allows a continuation of our current lifestyle. Even if the physics allows it, many other practical and economic barriers may limit options or successful implementations."

- "If we unwisely mount a response only after we find ourselves in fossil fuel decline—as crisis responders, not proactive mitigators—we could find ourselves in an energy trap: a crash program to build a new energy infrastructure requires up-front energy, for decades. If energy is already in short supply, additional precious energy must be diverted to the project, making people's lives seem even harder/worse. A democracy will have a hard time navigating this decades-long sacrifice."
- "We lack a plan for how to live within planetary limits."

He came to these results not from ideology or preconceptions, but from science that contradicted his preconception that human ingenuity would solve the problems as we'd solved problems historically.

Next: **Donella Meadows, Jorgen Randers, and Dennis Meadows**'s **2004 book** *Limits to Growth, the 30-Year Update,* which I consider the science book of the 2000s decade, combined with peer-reviewed comparison against fifty years of data showing remarkable correlation with the book's simulations.[86] The 1972 version sold ten million copies and prompted angry criticism. I've looked up much of that criticism and have found no example that understands the book. The authors state:

> We do not write this book in order to publish a forecast about what will actually happen in the twenty-first century. We are not predicting that a particular future will take place. We are simply presenting a range of alternative scenarios: literally, 10 different pictures of how the twenty-first century may evolve. We do this to encourage your learning, reflection, and personal choice.

Limits to Growth projects how the human economy interacts with renewable resources, nonrenewable resources, pollution, and variables to affect our population level, quality of life, longevity, and other measures

---

86 Herrington, "Update to Limits to Growth."

of well-being. It includes feedback and delays, setting their analysis nearly as far from Malthus as a cell phone from Morse code. They ran multiple simulations to explore parameter space to inform what humanity can in principle choose among priorities. For example, based on evidence, how much should we spend on developing technology to reduce pollution versus on industrial capital, protecting soil, and other outcomes?

Unknowns in inputs and assumptions led them to make many simulations. The critics seem to have interpreted simulations as predictions and concluded, "We haven't collapsed so *Limits to Growth* is wrong." Some saw the assumptions as overly simplified. As far as I've found, few critics have criticized the book's actual content (I suspect many haven't read it), and those I've found who have, I haven't found compelling.

The book shows patterns of how the variables they track interrelate and how our choices affect them. Before *Limits to Growth* people just threw up their hands at complex questions like: Can we solve a lack of clean water by desalinating? The answer is complex because desalination affects many other things. For example, it requires energy and affects the environment. Energy we use to desalinate doesn't go to other parts of the economy, meaning it affects business and government. Meanwhile, changing the environment may lower its ability to sustain life, requiring more resources to fix it or accepting a lower human population, quality of life, or both.

Limits to Growth showed how these variables interact and the potential results of human behavior. In this regard, the book helps leaders. To read it as a prediction misses or distorts the book's point.

I include Herrington's comparison of their simulations with fifty years of data. She found that "empirical data showed a relatively close fit for most of the variables" for two of the simulations. A close fit for ten variables representing disparate parts of the world with no reason to correlate over half a century lends confidence to the model.

Next, **David MacKay's 2009 book** *Sustainable Energy—Without the Hot Air* (a free download), which I consider the science book of the 2010s decade. You may recall MacKay's name from his video prompting me to experiment avoiding flying. He was a physicist and engineer at Cambridge University, trained at Caltech, and became Chief Scientific Advisor to the UK's Department of Energy and Climate Change.

He described the book's origin:

> I recently read two books, one by a physicist, and one by an economist. In *Out of Gas*, Caltech physicist David Goodstein describes an impending energy crisis brought on by The End of the Age of Oil. This crisis is coming soon, he predicts: the crisis will bite, not when the last drop of oil is extracted, but when oil extraction can't meet demand—perhaps as soon as 2015 or 2025. Moreover, even if we magically switched all our energy-guzzling to nuclear power right away, Goodstein says, the oil crisis would simply be replaced by a nuclear crisis in just twenty years or so, as uranium reserves also became depleted.

> In *The Skeptical Environmentalist*, Bjørn Lomborg paints a completely different picture. "Everything is fine." Indeed, "Everything is getting better." Furthermore, "We are not headed for a major energy crisis," and "There is plenty of energy."

> How could two smart people come to such different conclusions? I had to get to the bottom of this.

This accessible book helps you answer that question and assess humanity's ability to maintain our style of life.

Next: **Vaclav Smil's** *How the World Really Works: A Scientist's Guide to Our Past, Present and Future* (2022). Smil is a professor emeritus in the Faculty of Environment at the University of Manitoba. Of his over forty books (and counting), this one is the most popular and accessible, with *Numbers Don't Lie: 71 Stories to Help Us Understand the Modern World* a close second. He recognizes that "growth must come to an end. Our economist friends don't seem to realise that." In a *TIME* article he summarized a top point of this book:

Four materials rank highest on the scale of necessity, forming what I have called the four pillars of modern civilization: cement, steel, plastics, and ammonia . . . But it is ammonia that deserves the top position as our most important material: its synthesis is the basis of all nitrogen fertilizers, and without their applications it would be impossible to feed, at current levels, nearly half of today's nearly 8 billion people . . .

Fossil fuels remain indispensable for producing all of these materials.

He shows how nuclear, fusion, solar, wind, and other sources can't substitute for what we've become dependent on fossil fuels for, among other results. He also experimented living more sustainably.

Some articles summarizing the above include "The Green Growth Delusion" by **Christopher Ketcham** 🎤 and "Don't Call Me a Pessimist on Climate Change. I Am a Realist" by **William Rees**.[87, 88]

Next: **James Suzman**'s *Affluence Without Abundance* (2017). As important to finding a path to sustainability is understanding that we will prefer a sustainable world. I was pleasantly surprised to find anthropology provided it.

Suzman is an anthropologist who lived among the San Bushmen in southern Africa. This book describes their lifestyle, still hanging on to what Sustainable Free Abundant culture it can, and how PAID culture is destroying it. He said of his book:

Only in the last 10 years or so have we begun to understand just how ancient the Bushmen are, and quite how enduring that culture is. New archaeological data and genomic data have revealed that the Bushmen were extraordinarily isolated from other groups, and in particular from modernity and the agricultural revolution. They've been around an astonishingly long amount of time, and most likely

---

87 Christopher Ketcham, "The Green Growth Delusion," *Truthdig*, April 4, 2023.

88 William E. Rees, "Don't Call Me a Pessimist on Climate Change. I Am a Realist," *Resilience*, November 12, 2019.

lived in a similar manner for a period stretching back 70,000 years, possibly longer. This may give us pretty good insight into how homo sapiens lived for 95 or 98 percent of human history . . .

If we judge a civilization's success by its endurance over time, then the Bushmen are the most successful society in human history. Their experience of modernity offers insight into many aspects of our lives, and clues as to how we might address some big sustainability questions for the future.

On human impact on nature, **J. B. MacKinnon**'s 🎙 *Once and Future World* (2014) clarifies that while humans can't avoid affecting nature and nature has always changed, we've dramatically reduced earth's ability to sustain life. He wrote of the "sheer abundance of life" recorded in the past "is an astonishment. In the North Atlantic, a school of cod stalls a tall ship mid-ocean; off Sydney, Australia, a ship's captain sails from noon until sunset through pods of sperm whales as far as the eye can see . . . When the shad pulse up the Hudson River to spawn, they push a wave like a tidal bore in front of them." The book is full of such astonishments we don't know we're missing and can restore.

On anthropology, I recommend the resources I mentioned throughout the book: **Sebastian Junger**'s *Tribe* and the *What Is Politics?* videocast by **Daniel** 🎙 (who goes by his first name only), both creators with degrees in anthropology. [89]

On population and overpopulation, I recommend **Jane O'Sullivan**'s 🎙 research in overpopulation.[90] O'Sullivan is Honorary Senior Fellow in the School of Agriculture and Food Sustainability in the Faculty of Science at the University of Queensland, Australia. Her research shows the flaws in widespread belief that population growth in today's conditions creates jobs

---

89 https://www.youtube.com/@WHATISPOLITICS69

90 Joshua Spodek, "Jane O'Sullivan videos on population and overpopulation," December 7, 2023.

or prosperity, why people believe it does, and what would work instead. I recommend starting with her videos.

I couldn't talk about population until I read **Alan Weisman**'s 🎤 2013 book *Countdown: Our Last, Best Hope for a Future on Earth?*, where I learned of many nations that managed population with the opposite of China's One Child Policy, the opposite of eugenics or racism. Policies that work are voluntary, non-coercive, even fun, and he gives many examples. If you still believe that acting on population means eugenics, racism, Nazism, forced abortion, forced sterilization, authoritarianism, or that it conflicts with your beliefs, it will broaden your horizons.

Marc Reisner's *Cadillac Desert: The American West and Its Disappearing Water* documents booms followed by bust, dependence, plunder, and socialism for the rich. They resulted from acting on overly hopeful and optimistic but baseless claims of overriding natural processes taking population growth for granted and disregarding long-term unintended side-effects. This book documents the case of believing "the rain follows the plow." Believing "more people solve more problems" seems likely to follow similar patterns.

Next: Two free online engineering resources: *Low Tech Magazine* by **Kris de Decker** 🎤 promotes a view of technology and innovation for overall effects on culture and life, not just intended effects. I read the article that prompted me to unplug my refrigerator there.[91] The *Not Just Bikes* video series by **Jason Slaughter** 🎤 and **Strong Towns** started by **Charles Marohn** 🎤 similarly look at civil engineering and city planning.[92,93]

Results: The Intuitions I've Developed

I'm about to share how nearly all of the proposed solutions you've heard of step on the gas, while we think its the brake, including fusion, solar, wind, and carbon taxes. I avoided researching them most of my life, fearing these

---

91 Aaron Vansintjan, "Vietnam's Low-tech Food System Takes Advantage of Decay," *Low Tech Magazine*, February 20, 2017.

92 https://www.youtube.com/@NotJustBikes

93 https://www.strongtowns.org

results and losing hope. I could only start researching when my experiments revealed that PAID culture's solutions weren't necessary for thriving either as an individual or as a culture.

I suspect these results will sound scary if you come to them before you act. Hope is nice, but testing is important. If something can't work, misplaced hope delays action.

Case Study: Norman Borlaug and the Green Revolution

Norman Borlaug is often called the "father of the Green Revolution." He may top the list of people who used technology, market incentives, and laws to alleviate suffering. Seeing people dying of hunger, he went out in the fields, worked tirelessly and painstakingly for years, and developed agricultural technology to increase yields beyond anyone's expectations. He is credited for saving over a billion lives, earning him the Nobel Prize, the Presidential Medal of Freedom, and the Congressional Gold Medal.

By the time he was awarded the Nobel Prize, though, he saw effects undoing his work intended to save lives. In his Nobel acceptance speech, he said:

> The Green Revolution has won a temporary success in man's war against hunger and deprivation; it has given man a breathing space. If fully implemented, the revolution can provide sufficient food for sustenance during the next three decades. But the frightening power of human reproduction must also be curbed; otherwise the success of the Green Revolution will be ephemeral only.
>
> Most people still fail to comprehend the magnitude and menace of the "Population Monster" . . . Since man is potentially a rational being, however, I am confident that within the next two decades he will recognize the self-destructive course he steers along the road of irresponsible population growth.

Practical experience taught him sustainability science, especially systems with feedback and delays. His point wasn't that he couldn't imagine people forever solving ways to create more food. He wasn't repeating Malthus. He saw that the mechanisms of technology, market incentives, and legislation helped hungry people but *accelerated and grew the system that caused the hunger*. He knew these mechanisms (technology, market incentives, and law) better than anyone, but recognized he was stepping on the gas, thinking it was the brake. He repeated this warning for decades. He acted on it by joining the board of the Population Media Center, which helps limit population through voluntary non-coercive ways, now led by Bill Ryerson 🎙.

Another historical example of efficiency causing more consumption is James Watt's steam engine at the dawn of the Industrial Revolution. His engine was more efficient than any before so people expected coal use to drop. It didn't. While each engine used less coal per use, more people used more engines for more uses, so total coal consumption increased.

When Technology, Market Incentives, and Laws Backfire

Another example is Eli Whitney's cotton gin. It was designed to require less labor, so one might have expected it to reduce slavery. Plantation owners who bought them didn't value less labor, they valued more output, so it helped create the largest slave empire in history.

Another: Robert Moses built highways around New York City to reduce congestion, expecting more roads to give drivers more space. Instead they prompted more car use, resulting in more congestion. His practices were replicated around the world.

Another: The inventors of heroin created a more pure form of morphine. It was marketed to reduce addiction. As we know, it led to more addiction. The practice of purifying something addictive has been repeated with cocaine, crack, nicotine vaping, and doof.

Other examples of similar systemic effects leading to "stepping on the gas" results: making car engines more efficient led to making the cars bigger and to people driving them more, leading to more pollution and depletion; more refrigeration has led to longer supply chains and diets with less fresh produce; airplanes are more efficient than ever and the lower costs are leading to more flights; plastic was invented to replace ivory in billiard balls and reduce elephant hunting, but has contributed to technology and an economy shrinking elephants' territory, accelerating their demise; electric lighting reduced the demand for whale oil but accelerated market cycles that led to reducing whale populations. The individual cases aren't anomalous: our global economic system is more efficient than ever and polluting and depleting more than ever.

Economists have names for the pattern ("Jevons paradox" and "rebound effects"), but I call it "pulling an Eli Whitney." When you understand systems, you see it's not a side-effect we can innovate out of. It's how systems work. Making a system more efficient may lower waste in one area, but it will accelerate the system overall. In other words, **if you make a polluting, depleting, plundering system more efficient, you pollute, deplete, and plunder more efficiently**.

One electric vehicle may emit less pollution than *one* gas-powered car, but our culture is pushing *more* cars, *more* roads, *more* PAID culture. Tesla is pulling an Eli Whitney. Fusion would pull the greatest Eli Whitney in history—that is, it would accelerate PAID culture.

Since polluting and depleting destroy others' life, liberty, property, and freedom, **if you make a system that destroys others' life, liberty, property, and freedom and plunders more efficient, you destroy others' life, liberty, property, and freedom and plunder more efficiently**.

Technology, market incentives, and laws are tools. Like sharp knives and fire, they augment the values of those using them. No matter the intent of the innovator or people using them, culture will bend them to its purposes. As they say in business, "Culture eats strategy for breakfast."

I'm not saying we shouldn't pursue technology, market incentives, and laws. I'm saying we shouldn't pursue them in a coercive, tyrannical culture. First make our markets and culture free and not coercive—which the APPLE PIE Amendment helps achieve—then these mechanisms will accelerate that system: **if you make a system that creates freedom more efficient, you create freedom more efficiently**.

Carbon Offsets

Intuition: *Carbon* emissions *are a problem, but the real problem is* new carbon introduced to the biosphere, *as from extracting fossil fuels.*

Last June, smoke from wildfires in Canada made my neighborhood look like this:

I had seen pictures more severe from other fires around the world for years. From these fires, "2bn tonnes (2.2bn tons) of carbon dioxide were released," over triple the amount the rest of Canada's economy releases in total.[94]

Lightning has caused fires since long before humans, yet CO2 levels

---

94 Leyland Cecco, "Wildfires Turn Canada's Vast Forests from Carbon Sink into Super-Emitter," *The Guardian*, September 22, 202393

in the atmosphere have been nearly constant since before civilization and more stable for millions of years than they are in my lifetime.[95] Why haven't millions of years of forest fires caused global warming? Because they released carbon already in the biosphere. Burning things in the biosphere shuffles carbon around.

Extracting fossil fuels from underground is different. It brings new greenhouse gases into the biosphere. Carbon in fossil fuels has been outside the biosphere for tens to hundreds of millions of years. As far as humans are concerned, it is new carbon.

Measuring carbon *emissions* misses the problem of new carbon. A relevant, meaningful measure is how much *new carbon is introduced to the biosphere.*

Why Carbon Offsets Pull an Eli Whitney

You may have read of investigations of carbon sequestering schemes that reported fraud and that these schemes are ineffective and often lead to more emissions.[96] Those schemes are a problem, but the investigations miss that the concept fails by design.

Carbon offsets offset *emissions*. If you offset *emissions* from fossil fuels with sequestration elsewhere in the biosphere, you're paying to *introduce more carbon into the biosphere*, increasing greenhouse concentration. We are developing technologies to sequester carbon, but in negligible amounts relative to fossil fuel extraction rates, with negligible ability to remove it from the biosphere.

In short, carbon offsets fund bringing new carbon into the biosphere.

Why Carbon Taxes Fail

Intuition: *Carbon taxes are a category error and moral hazard.*

---

95 "Global CO2 Levels," CO2levels.org

96 Patrick Greenfield, "Revealed: More Than 90 Percent of Rainforest Carbon Offsets by Biggest Certifier Are Worthless, Analysis Shows," *The Guardian*, January 18, 2023.

Proponents of carbon taxes suggest they would discourage emissions and, if returned through a dividend, help the working class. Besides focusing on emissions instead of new carbon, they cause government to generate revenue from emission and extraction—activities destroying life, liberty, property, and freedom and plundering. They motivate government to grow the practices, its size, its power, and corruption.

It's tempting to suggest that people should pay for their externalities, which a tax could do, but pollution, depletion, and plunder aren't externalities. Again, we *tax* selling cigarettes to adults and don't limit them from smoking them at home, but we *make illegal* selling them to minors and smoking them where people can't avoid their second-hand smoke

Why Carbon Capture Pulls an Eli Whitney (Or Would If It Worked at Scale)

Intuition: *Bringing fossil fuels into the biosphere necessarily destroys life, liberty, property, and freedom.*

It's tempting to say, "You're underestimating innovation. Capture and sequestration could become huge." Fossil fuel emissions poison in ways besides carbon. Extracting displaces people from their lands. Even if we could sequester and store carbon on geological time scales, we'd have to offset the new carbon we've introduced for centuries before it could offset new extraction. Even then, we could only use it to offset carbon emissions, not other pollution. It wouldn't restore refugees' homelands.

It's tempting to say, "Burning fossil fuels helps people. We have to balance the good with the bad." Monticello is a beautiful building if you ignore that it resulted from slavery, but we don't balance plantation mansions' beauty with slavery.

I tried to find how much oil we could use without harming others. Here is an industry breakdown of uses of oil. Can you find any outputs that, for whatever benefit they provide one person, they do not also destroy someone else's life, liberty, property, and freedom without consent?

CRUDE OIL
FRACTIONING COLUMN

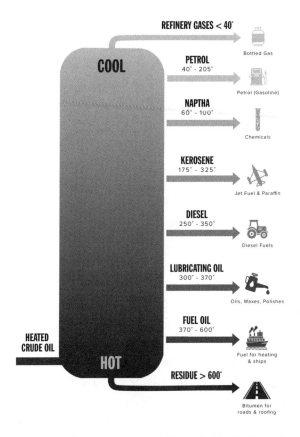

REFINERY GASES < 40°

Bottled Gas

COOL

PETROL
40° - 205°

Petrol (Gasoline)

NAPTHA
60° - 100°

Chemicals

KEROSENE
175° - 325°

Jet Fuel & Paraffin

DIESEL
250° - 350°

Diesel Fuels

LUBRICATING OIL
300° - 370°

Oils, Waxes, Polishes

FUEL OIL
370° - 600°

Fuel for heating
& ships

HEATED
CRUDE OIL

HOT

RESIDUE > 600°

Bitumen for
roads & roofing

Why Today's Solar and Wind Fail in the Long Run

Intuition: *Solar and wind are renewable but the tools to date to harness them pollute and deplete on par with fossil fuels.*

Windmills, solar panels, and batteries require fossil fuels for manufacture, transportation, installation, and decommissioning. Extracting the materials they require will affect the environment and people as much as using fossil fuels. They don't enable long-distance flying, trucking, or ocean shipping, fertilizer, making steel, or other areas our culture demands.

Some exceptions: Solar energy harnessed by plants through photosynthesis is beyond sustainable; it's regenerative. Passive solar heating, windmills like Don Quixote fought, sailboats, and other traditional technologies work, though on small scales. I expect that innovation can help make solar sustainable, but continuing to use solar that relies on materials that require pollution, depletion, and plunder delays that innovation. I view solar and wind like methadone or suboxone—still powerful opiates, and if you give them to people addicted to stronger opiates who don't acknowledge their addiction or resolve to stop, they won't like them as much as the stronger stuff, but they'll use them *and* the stronger stuff. But if you give them as part of a plan to stop to people who acknowledge their addiction and intend to stop, they can help. It's tempting to think using more solar and wind is replacing fossil fuels, but PAID culture hasn't acknowledged its addiction nor resolved to live sustainably, so we use solar and wind *and* everything else.

Why Nuclear and Fusion Pull an Eli Whitney

Intuition: *Putting a piece of the Sun within our atmosphere will accelerate global warming.*

A tsunami caused a shutdown of a nuclear power plant in Fukushima, Japan, releasing radioactive contaminants, an incident rated the maximum severity, level seven. Fukushima is five hundred miles from Hiroshima. The Japanese know atomic dangers, tsunamis, and engineering as well as anyone. Why did they build a reactor by the sea in an earthquake-prone region where a tsunami could hit it?

Besides generating power, the plant generates heat, which needs to be dissipated, which large bodies of water do. When the water cools the plant, the plant heats the water. Plants fueled by fossil fuels or fusion will heat the environment too. So do most factories. One plant's heat is negligible on a global scale, but what about the combined heat from industry around the globe?

The heat from industry overall is small compared to global warming, but it's not zero. If we kept growing the economy, we'd have to cool more plants and factories, meaning we'd keep warming more ocean.

It turns out that if the global economy grew at historical rates (below 2.5 percent per year), "waste heat would rival global warming next century."[97] That is, if we tried to keep growing at that rate, within a lifetime we would double global warming by heating our atmosphere. We can't grow at that rate much longer. Even if we developed fusion that emitted no greenhouse gases or other pollution, had no risk of failure or terrorism, and had infinite fuel, we can't escape the waste heat from powering that growth. Now that you know about systems, you know we'd hit problems before overheating since accelerating our economy would accelerate pollution, depletion, and plunder in all other parts of the economy, like plastics, carcinogens, corruption, and so on.

Putting a piece of the Sun within our atmosphere will heat us faster than just trapping the Sun's heat. Any other source would heat us as quickly while polluting more too.

What About Other Technologies?

The media report on a bewildering array of potential solutions, with billions of dollars of investment and subsidies. Surely some will work.

Doing the math, engineering, and physics matter. Murphy's *Do the Math* blog and his textbook analyze these subjects from first principles. Some technologies solve some problems, but none enable PAID culture to continue. None stop PAID culture from plundering, destroying life, liberty, property, and freedom today.

Decoupling, Substituting, Dematerializing, and Efficiency Pull an Eli Whitney

It's tempting to suggest that by decoupling, substituting, dematerializing, and otherwise creating more efficiency, we could lower pollution and depletion to zero. They may lower pollution and depletion a bit, but not

---

97   Tom Murphy published the calculations in the peer-reviewed "Limits to Economic Growth," *Nature Physics.* He also published them more for the lay person in his textbook *Energy and Human Ambitions on a Finite Planet* and his blog, *Do the Math,* both described in this coda.

to zero, and these actions accelerate and grow pollution and depletion elsewhere in the economy.

It's tempting to dream all the waste will go to zero, but it can't. Even if the only waste were heat, in an exponentially growing economy, as mentioned earlier in the context of nuclear and fusion, our waste heat would grow exponentially. We produce heat by cooking food, transporting ourselves, creating electricity, and so on. Even if we could reduce our material existence to the minimum possible, downloading our minds onto silicon chips, we can't avoid producing heat. In practice, we also produce material waste like plastic and other things that fill landfills, which efficiency accelerates overall.

All those problems occur when creating energy combines with growth. What if we stop growing?

What If We Stop Growing?

If we stop growing our energy needs, we can solve problems of overheating. We can keep innovating and improving lives, which don't require more people or higher GDP. The APPLE PIE Amendment makes it possible to stop growing by removing the need to plunder resources from others.

If we believe we can stop growing in the future, it is easier to stop growing now. Even if we keep growing to Mars and even other galaxies, we'd have to stop growing on earth. If we think we can ever stop growing on earth, let's do it now.

Do we need to grow? Milton Friedman didn't oppose growth but clarified, "We have no desperate *need* to grow. We have a desperate *desire* to grow, and those are quite different. I believe that the level of growth in this country ought to be whatever people want it to be. If the people at large—if each and every person separately was satisfied with where he is and didn't want to grow, fine. I have no objection. I don't want to impose growth on anyone. I want people to be free to pursue their own objective." While he implied some benefits to growth, he also noted, "The bigger they are the harder they fall," then listed corporations larger than many nations that went bankrupt. Growth doesn't provide immunity from collapse.

It bears repeating that the immediate problem is not overheating later but coercion, plunder, and destroying freedom now. It's tempting to say we can solve Cancer Alley by finding a cure to cancer through growth, but even curing cancer wouldn't undo the suffering and destruction of life, liberty, and property *today*.

As for growth solving these problems, the problem was causing Cancer Alley in the first place. If people can legally profit while causing cancer or any other environmental problem, they will grow their profit and therefore those problems, pointing to their profit and wealth as evidence of having helped people.

That line of thinking is tempting, the opposite of the Enlightenment values of reason and science. Slaveholders similarly promoted "spread theory"—that growing slavery over more territory would end it sooner. But growing coercion and tyranny grow coercion and tyranny.

Population

Will our population level off? People point out that population growth is slowing. Does that mean we averted the problem? Haven't we proved Malthus and the doomsayers wrong? I find it helpful to distinguish between three independent issues of overpopulation: depletion—which Malthus treated—pollution, and complex systems involving feedback and delays, which he didn't.

I described complex systems in the section on *Limits to Growth* and fifty years of data showing remarkable correlation. Looking at depletion and pollution, we see that Malthus's ideas were a nice start, but not relevant to this book.

Depletion

I'll illustrate depletion, or running out of resources, with a conversation with Kevin Cahill 🎙. He told me of a time he spoke to a neighbor after buying land in Idaho. Kevin saw elk for the first time. The neighbor was a hunter who had lived there his whole life.

From the porch they saw a herd in the distance. The hunter asked, "How many elk do you see?"

"Oh, about a hundred."

"Yeah, I'd say about that. Over the years, I've learned the vegetation around here is enough to support about seventy-five over the winter."

Kevin thought a moment. "I guess that means twenty-five will die."

"No," the hunter paused for effect. "Twenty-five will *live*." Kevin paused to let me figure it out. The elk don't measure the vegetation and ask twenty-five not to eat so the others can live. No, as winter starts, all hundred eat. Soon there's enough food for seventy-four but still a hundred elk. All keep eating and soon there's enough for seventy-three, then seventy, fifty, forty, and so on but still a hundred elk. Then the food becomes too scarce and they start dying. Twenty-five eke it out, emaciated, to springtime.

The hunter may be biased, motivated to imply that hunting some elk saves more, so maybe he exaggerated, but he didn't make up the effect. Hunters know it. Gardeners know it too if they've seen a pest without predators eat a crop until none remains, and then die off themselves.

Most people don't realize the numbers. Overshooting what a necessary resource can sustain by small amounts can lead to dramatic drops.

Populations can collapse even to extinction, as happened on the remote Alaskan St. Matthew Island. In World War II, the coast guard operated a base there. Its long, brutal winters meant they might not be able to resupply the base, so they introduced to the island twenty-nine reindeer for the men to hunt for food if necessary. The men didn't end up needing to. The war ended. The men went home and left the reindeer.

Biologists visited in 1957 found the reindeer could handle the winters and thrived. They counted over a thousand healthy, fat reindeer. In 1963, they returned. "We counted 6,000 of them," one said, sensing an imbalance with their food source.[98] "They were really hammering the lichens." In 1966 they returned to find only forty-two alive. Reindeer skeletons were everywhere. The only male was infertile so when those forty-two died, a population with no predators that started with abundant food went extinct on the island.

You might say, "The Green Revolution proved Malthus wrong. We aren't animals. We can use technology, legislation, and market incentives to overcome limits animals can't." Even only considering the one environmental factor Malthus did—running out of resources—we still face problems.

It's tempting to say we'll use our intelligence to solve environmental problems, but countless human cultures have collapsed for environmental reasons. It's tempting to say we know better than those cultures, but they thought they knew better and didn't expect to collapse either. It's tempting to say technology will save us or that markets will make things expensive before we use them up to force us to stop using them, but technologies tend to pull Eli Whitneys.

Differences can work against us too. Elk don't say, "If you cross this line I will defend my territory" and start wars. Elks don't have nuclear weapons or pollute and deplete in ways that affect everyone, not just combatants.

Unlike animals, we can look at patterns and predict the future. We have empathy, compassion, and teamwork, so we have the potential to solve such problems. We have more than months to solve our problem. Still, even if our population levels off, if we're in overshoot (see below), our population will collapse, likely exacerbated by war. Unlike the elk surviving winter, no springtime will bring us new nonrenewable resources.

---

98   Ned Rozell, "Bitter Weather May Have Wiped Out Reindeer," Geophysical Institute, January 7, 2010.

Pollution

Malthus lived before pollution like today's so didn't consider it. I'll illustrate how pollution affects population with the yeast that makes wine. You may have noticed wine's alcohol content peaks around fifteen percent. That's because alcohol is yeast's poop and it kills the yeast. Wines with the highest alcohol content tend to be sweeter because making yeast grow in its poop requires giving it energy—in this case, sugar.

We enable ourselves to live in overshoot the way we enable yeast: by giving ourselves energy—in our case, fossil fuels, nuclear, and so on—which degrade our environment, beyond what Malthus considered.

Resolving the UN's Confusion About Population

Remember the UN projections?

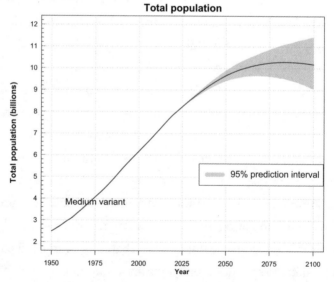

It's tempting to think our population will level off, but we've seen that driving growth with extra energy results in what people who work with systems call "overshoot and collapse" and it diverges from UN projections.

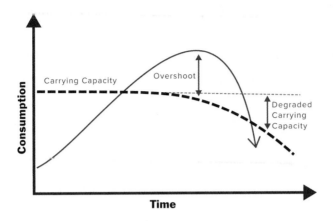

The effect occurs when a population overshoots what its environment can sustain—its *carrying capacity*. Living in overshoot lowers the carrying capacity, so the population ends up falling well below what could have survived, hence seventy-five elk dying and all the reindeer.

How would humans fare in a collapse? Nobody can say for sure and we won't get a second chance, but we can estimate in many ways. Since the Green Revolution requires nonrenewable resources and isn't sustainable, many estimate that if humanity overall impacts the environment at Western European-levels of impact, earth could sustain about the population before the Green Revolution: three billion people.[99] If so, eight billion puts us into overshoot by over a factor of two, suggesting collapse would bring our population to a fraction of three billion.

It's hard to think about such a result. I'm not promoting or demoting this estimate, but many scientific predictions seemed unbelievable until they happened. What was Galileo supposed to do after seeing moons orbiting another planet but conclude the earth wasn't at the center of the universe? What was Einstein supposed to do about experiments that showed the speed of light had to be constant, other than conclude that it was?

---

99   Christopher Tucker, *A Planet of 3 Billion* (2019).

We do ourselves greater harm clinging to visions no longer possible than to accept what we see and move forward. People who accepted and embraced Galileo's and Einstein's results helped create brighter futures. What better option do we have than to accept what observation shows us, embrace it, and make the brightest future we can from it, even if people clinging to visions no longer possible call it ghoulish and eco-pessimist?

At last we can resolve the UN's confusing, conflicted message from the opening of this book. One part of the UN screeched "Code Red!" while another part presented the population smoothly leveling off, even decreasing. Which is it?

The UN's assumptions make it impossible for its projections to show population dropping fast. As I understand its models, if applied to cultures that did collapse from foreseeable environmental problems, it wouldn't be able to predict them.

The projections are based mostly on demographic information: nation-by-nation profiles in age, sex, education level, immigration, and related data. The UN publishes its methodology.[100] It doesn't allow that resources may become depleted, only new estimates of birth, death, and migration. At its core,

> The preparation of each new revision of the official population estimates and projections of the United Nations involves two distinct processes: (a) the incorporation of new information about the demography of each country or area of the world, involving a reassessment of past estimates where warranted; and (b) the formulation of detailed assumptions about the future paths of fertility, mortality and international migration, again for every country or area of the world.

Their projections include non-demographic information for "Special

---

100   UN Department of Economic and Social Affairs Population Division: Methodology Report: World Population Prospects 2022.

considerations for countries highly affected by HIV and AIDS," for COVID-19, and "Average relative risks of mortality by age and sex were estimated for nine categories of crisis events (battle deaths, conflicts, genocide (including mass killings), cyclones, earthquakes, epidemics, famine/droughts, floods, Tsunami)." Based on estimations, they adjust the uncertainty on the graph—the shaded regions above and below the main projection.

Researchers at Cornell predicted, "In the year 2100, 2 billion people—about one-fifth of the world's population—could become climate change refugees."[101] I would expect such an outcome to affect the population beyond what the UN assumptions would allow. Likewise, a volcano or earthquake may be unlikely in any given year, but rivers and aquifers supporting millions of people running dry, fisheries collapsing, and bio-diversity decreasing aren't "special considerations." They are happening and ongoing. If anything would be "special" in such situations, it would be solutions to these problems, but the problems persist.

I don't find the UN projections of stability and slow decline credible. When you include feedback loops with human impact on the environment, it's hard to avoid population dropping faster than our economies and culture could respond to avoid major disruptions. The assumptions I know of that prevent rapid population decline are ones including widespread cultural change restoring stewardship and sustainability. The Spodek Method creating a movement leading to an APPLE PIE Amendment provides such a change, as far as I know uniquely, which is why I practice and teach it.

Overshoot in Art

It turns out a poem describes overshoot and collapse—Percy Shelley's "Ozymandias":

---

101 "Impediments to inland resettlement under conditions of accelerated sea level rise," as reported in *Cornell Chronicle*.

I met a traveler from an antique land,
Who said—"Two vast and trunkless legs of stone
Stand in the desert. . . . Near them, on the sand,
Half sunk a shattered visage lies, whose frown99
And wrinkled lip, and sneer of cold command,
Tell that its sculptor well those passions read
Which yet survive, stamped on these lifeless things,
The hand that mocked them, and the heart that fed;
And on the pedestal, these words appear:
My name is Ozymandias, King of Kings;
Look on my Works, ye Mighty, and despair!
Nothing beside remains. Round the decay
Of that colossal Wreck, boundless and bare
The lone and level sands stretch far away."

"Look upon my Works, ye Mighty, and despair" expresses the hubris that ecomodernists and techno-optimists do, suggesting that past results indicate future performance and not to question their visions. The people in these movements think they're different, as Ozymandias did.

It may help to review a second "Ozymandias" poem, by Shelley's friend Horace Smith:

In Egypt's sandy silence, all alone,
Stands a gigantic Leg, which far off throws
The only shadow that the Desert knows:—
"I am great OZYMANDIAS," saith the stone,
"The King of Kings; this mighty City shows
The wonders of my hand."—The City's gone,—
Naught but the Leg remaining to disclose
The site of this forgotten Babylon.
We wonder—and some Hunter may express
Wonder like ours, when thro' the wilderness
Where London stood, holding the Wolf in chace,

He meets some fragment huge, and stops to guess
What powerful but unrecorded race
Once dwelt in that annihilated place.

The poems mention Egypt and potentially London as sites of overshoot and collapse. If you've visited Machu Picchu, Angkor Wat, or innumerable other sites of lost cultures, you've seen more cultures have collapsed than the one PAID culture dominating and destroying those that remain.

Yet Sustainable Free Abundant cultures have been fruitful and multiplying for periods up to thousands of times longer than the time period stretching from the Enlightenment to today. PAID culture appears on track to become an Ozymandias-like blip in the histories of these cultures that survive. Instead of leaving a desert behind, we would leave carcinogens, endocrine disruptors, and greenhouse gases that would leave cancer, birth defects, and uninhabitable regions for millennia.

We can instead endure as Sustainable Free Abundant cultures have. We don't have to become just like them, though some might choose to. I only propose that we restore a value we once practiced and which they still do—stewardship—and use it to unify our house divided and make our union more perfect. I believe ecomodernists and peers will come to embrace the APPLE PIE Amendment, though after gut checks.

What I can say that nearly no one else can, for my experience leading others and myself to live more joyfully sustainably, is that when you live it for your reasons, anyone who tries to return you to PAID culture will have to tie you up like those women forced to return to colonialism. You will need no willpower to keep living joyfully sustainably or to resist doof, isolation, and addiction.

Workshop participants and Spodek Method practitioners can tell you more, that by practicing with a few friends or a workshop cohort, a transition that took me years you will be able to do in weeks. You can download the how-to workbook free at SpodekMethod.com and start today.

You'll wish you started earlier.

Illustration Credits

Page 24 and page 308: ©2022 United Nations, DESA, Population Division.

Licensed under Creative Commons license CC BY 3.0 IGO. United Nations, DESA, Population Division. *World Population Prospects* 2022. http://population.un.org/wpp/

Page 102: photo available in the public domain: https://pxhere.com/en/photo/1180874

Page 119: adapted from Dr. Anna Lembke's TEDx Talk, *Dopamine Nation: Finding Balance in the Age of Indulgence with Anna Lembke*, www.youtube.com/watch?v=n2u8Z1HeKD8&t=233s

Photos on pages 277 and 278 used with permission of Benenson Productions.

All illustrations and photos not mentioned here are from the author.